Investigations into Assessment in Mathematics Education

Published under the auspices of The International Commission on Mathematical Instruction under the general editorship of

Miguel de Guzmán, President   Mogens Niss, Secreta

Investigations into Assessment in Mathematics Education

An ICMI Study

edited by

Mogens Niss

Roskilde University, Denmark

Kluwer Academic Publishers

Dordrecht / Boston / London

ISBN 0-7923-2095-6

Published by Kluwer Academic Publishers,
P.O. Box 17, 3300 AA Dordrecht, The Netherlands.

Kluwer Academic Publishers incorporates
the publishing programmes of
D. Reidel, Martinus Nijhoff, Dr W. Junk and MTP Press.

Sold and distributed in the U.S.A. and Canada
by Kluwer Academic Publishers,
101 Philip Drive, Norwell, MA 02061, U.S.A.

In all other countries, sold and distributed
by Kluwer Academic Publishers Group,
P.O. Box 322, 3300 AH Dordrecht, The Netherlands.

Printed on acid-free paper

Printed in the Netherlands

TABLE OF CONTENTS

MOGENS NISS

ASSESSMENT IN MATHEMATICS EDUCATION AND ITS EFFECTS: AN INTRODUCTION

1. THE PRESENT BOOK AS AN ICMI STUDY

The present book, *Investigations into Assessment in Mathematics Education*, is one of two studies resulting from an *ICMI* Study Conference on **Assessment in Mathematics Education and Its Effects.** The book which is published in the series of **ICMI Studies** under the general editorship of the President and Secretary of ICMI is closely related to another study resulting from the same conference: *Cases of Assessment in Mathematics Education* (Niss, 1992). The two books, although originating from the same sources and having the same editor, emphasize different aspects of assessment in mathematics education and can be read independently of one another. While the present book is devoted to critically analyzing general and principal aspects of assessment, the other study presents and discusses cases of assessment that have been implemented in actual practice. Naturally, the content of either book is enriched by the materials and perspectives provided by the other one.

In order to put this book and its background into context, the nature and scope of the ICMI studies are outlined briefly below.

Since 1986 the *International Commission on Mathematical Instruction (ICMI)* has been engaged in publishing a series of studies on essential topics and key issues in mathematics education. Previously, the following studies have been published (all by Cambridge University Press): *School Mathematics in the 1990s* (Howson & Wilson, 1986), *The Influence of Computers and Informatics on Mathematics and Its Teaching* (Churchhouse et al., 1986), *Mathematics as a Service Subject* (Howson, et al., 1988), *The Popularization of Mathematics* (Howson, et al. 1990), *Mathematics and Cognition: A Research Synthesis by the International Group for the Psychology of Mathematics Education* (Nesher & Kilpatrick, 1990).

Depending on the theme under consideration a study may either be *research* oriented or *action* oriented (or both). In either case the aim is to provide an up-to-date presentation and analysis of the state-of-the-art concerning a theme, whether by identifying and describing current research contributions and their findings, or by identifying and discussing crucial, non-rhetorical issues involving genuine controversies or dilemmas and the different positions towards them held by various mathematics educators.

M. Niss (ed.), Investigations into Assessment in Mathematics Education, 1–30.

In order to provide a platform for producing an ICMI study the following normal procedure has been adopted (the exception is the study on cognition). The Executive Committee of ICMI appoints a fairly small, international *Program Committee.* Its first task is to write a so-called *Discussion Document* that outlines the theme, the aims, and the scope of the study, and presents the items and issues to be dealt with. The Discussion Document is published in international journals (including the official organ of ICMI, *L'Enseignement mathématique*) and newsletters with an invitation to mathematics educators to respond to the Document and to apply for participation in a so-called *Study Conference*.

The Study Conference is held with a limited number (50–100) of individuals and constitutes a working forum of experts and novices with ideas, experiences and expertise to investigate the theme of the study. This investigation is guided by the Discussion Document, assisted by working papers (written by participants), presentations, debates, and group work.

Finally, the study proper is produced and published under the general editorship of the President and the Secretary of ICMI, and based on the written materials and the work done at the Study Conference. As every study is written and edited as an independent publication for a wide international readership, its nature is *not* that of a conference proceedings.

In May 1989 the Executive Committee of ICMI appointed the following international *Program Committee*:

Claudi Alsina, local organizer, Universitat Politècnica de Catalunya, Barcelona, Spain;
Desmond Broomes, University of the West Indies, Bridgetown, Barbados;
Hugh Burkhardt, Shell Centre for Mathematical Education, University of Nottingham, UK;
Mogens Niss, chairman of the Program Committee, Roskilde University, Denmark;
Thomas A. Romberg, National Center for Research in Mathematical Sciences Education, University of Wisconsin-Madison, USA;
David Robitaille, University of British Columbia, Vancouver, Canada;
Júlianna Szendrei, O.P.I. (National Institute of Education), Budapest, Hungary.

The Discussion Document was officially published in *L'Enseignement mathématique*, **36**, fasc. 1–2, Janvier-juin 1990, 197–206, as well as in a number of other journals and newsletters.

The *Study Conference* which was held at Cap Roig, Calonge, Spain, 11–16 April 1991, had 80 contributing participants from 25 different countries in Europe, North America and the Caribbean, Asia, Oceania, Africa and the Middle East.

A Note on Terminology

To begin with some terminological clarification may be in order. The field we shall be dealing with frequently uses terms such as assessment, evaluation, tests, exams. However, these words and their counterparts in other languages carry quite different connotations within different educational systems and contexts. The variation is so large that the same word often has different meanings to different people. We shall confine ourselves to making one distinction, namely between *assessment* and *evaluation*. Further terminological questions will be addressed in later sections of this chapter.

While assessment and evaluation are often used interchangeably we shall adopt, as was suggested in the Discussion Document for the present study, the following terminological convention: **Assessment** in mathematics education is taken to concern the judging of the mathematical capability, performance, and achievement — all three notions to be taken in their broadest sense — of **students** whether as individuals or in groups, with the notion of "student" ranging from Kindergarten pupils to Ph.D. students. Assessment thus addresses the outcome of mathematics teaching at the *student level*. **Evaluation** in mathematics education is taken to be the judging of **educational or instructional systems**, in its entirety or in parts, as far as mathematics teaching is concerned. Evaluation may concern system components such as curricula, programs, teachers, teacher training, and specific segments of the educational system such as schools or school districts etc. So, evaluation addresses mathematics education at the *systems level*.

When tests and exams are considered to be ways of judging student performance they are special forms of assessment and are thus subsumed under the assessment category. As a contradistinction, when tests and exams are viewed as being part of the modes of operation of an educational (sub)system, or when the outcomes of tests and exams are used as indicators of the quality of such a system, as is the case with international performance comparisons, exams and tests belong to the realm of evaluation. This duality shows features of the general relationship between assessment and evaluation: Assessment items — in particular assessment results, but also assessment modes — may be involved in the judging of system aspects, hence they would form part of an evaluation activity. The converse normally will not hold for evaluation; for instance the appraisal of teachers will often involve a multitude of components having nothing to do with assessment of students. So, the relationship between assessment and evaluation is not a symmetrical one.

In the present study the emphasis will be on *assessment* as defined above rather than on evaluation. Due to the duality just mentioned this does not imply that evaluation issues will not be considered. However, only those aspects of evaluation which have to do with assessment of students will be given explicit attention.

Why a Study on Assessment?

In recent years, assessment has attracted increased attention from the international mathematics education community. There are numerous reasons for this. One seems to predominate. During the last couple of decades, the field of mathematics education has developed considerably in the area of ideals and goals, and theory and practice, whereas assessment concepts and practices have not developed so much.

The mathematics curriculum has claimed new territory. Firstly, when it comes to *content*, aspects of applications and modeling, cooperation with other subjects on topics of common interest, philosophy and history of mathematics, problem-oriented creativity, explorations and experiments aided by computers and informatics, have been included in quite a few programs and curricula round the world. Secondly, we have witnessed a remarkable expansion of the spectrum of *working forms* and student *activities*. Extended investigations of pure and applied mathematics, project work, scientific enquiry and debate, out-of-classroom activities, experimentation, group work, and so forth, are no longer utopian entities in mathematics teaching. As a result, a much broader notion of mathematics and mathematics education has emerged.

These developments have *not*, however, been matched by a parallel developments in assessment, where values, notion, and theory, practice, modes, and procedures are concerned. Consequently, an increasing mismatch and tension between the state of mathematics education and current assessment practices are materializing. It may well be the case that the ideals and goals of mathematics education were *never* really in accordance with the assessment modes available to mathematics educators but, as in former times post-elementary mathematics education was offered only to a minority of children and youth, the problems created by the mismatch were, perhaps, less serious, or at least thought by mathematics educators to be less serious. At any rate, expanding the notions of mathematics and mathematics education has undoubtedly widened the gap between contemporary mathematics teaching and traditional assessment practices.

This gap has put assessment on the agendas of mathematics educators. In the interest of truth, it should be said that this is a rather new phenomenon. The development of mathematics teaching during the last three or four decades has emphasized curriculum reform — of different and sometimes even contradictory types, that is true — as the most important task. Concurrently mathematics education, as an academic field, has focused attention on the conditions for and processes involved in the learning of mathematics, in particular regarding the formation and acquisition of mathematical concepts. This largely left assessment out. Thus, it was viewed as a less important factor in mathematics education, a factor that in addition was "external" to mathematics education in several respects. To the extent assessment has attracted the attention of mathematics educators, it has often been due to uneasiness about its role and function. Traditional assessment modes, especially examinations and tests administered "from outside", have, in many cases, formed one of the factors that hindered or slowed down curriculum reform.

Now that curriculum reform *has* been, or is being carried out in many places, the situation has changed. The roles, functions, and effects of assessment in mathematics education should no longer be neglected; rather, they should become objects of investigation and examination for several reasons (see Commission Internationale de l'Enseignement Mathématique, 1990):

- The roles, functions, and effects of contemporary modes of assessment are neither clear, nor well understood.

- Current assessment modes and practices involve conflicting interests, divergent aims, and unintended or undesired side-effects. In particular, it is difficult to devise assessment modes which at the same time: (a) allow us to assess, in a valid and reliable way, the knowledge, insights, abilities, and skills related to the understanding and mastering of mathematics in its essential aspects; (b) provide genuine assistance to the individual learner in monitoring and improving his or her acquisition of mathematical insight and power; (c) help the individual teacher in monitoring and improving his or her teaching, guidance, supervision, and counselling; (d) assist curriculum planners and authorities, textbook authors, and in-service teacher trainers in adequately shaping the framework for mathematics instruction.

- The difficulties involved in devising and employing effective, harmonious assessment modes, free from serious internal and external problems, seem to be fundamental and universal in nature, and hence worthy of being dealt with from an international perspective.

2. THE PURPOSES OF ASSESSMENT

Most of the purposes of assessment in mathematics education are not unique to mathematics but pertain to assessment in any subject taught in the educational system. So, much of what is going to be said about the assessment of mathematical performance will hold equally well for the majority of these subjects.

As long as there have been students of mathematics, they have been assessed, in one form or another, to determine their benefits and achievements. Any person (whether a professional teacher, a relative, or a friend) attempting to teach a segment of mathematics to someone else will judge — maybe unconsciously, maybe tacitly, maybe explicitly, maybe through set tasks — the outcome of his or her teaching.

As long as mathematics has been taught in *institutions* (schools, universities, etc.) some form of reporting — formal or informal, oral or written — of the outcome of the teaching has been required by those who are in charge of, or pay for, the institution, but often by the individual student as well. Such reporting activity tends to require a basis on which the outcome of teaching is to be judged and communicated. This has tended to stimulate the creation of certain more or less well-defined, and more or less standardized modes of assessment.

As long as competence in mathematics has been associated with the obtaining of positions, promotion, privileges, licences, admission to further schooling, or with the mastering of certain jobs and functions, the reporting of students' mathematical achievement has also served the needs of recipients outside of the institutions that take care of the mathematical education of students.

As long as there have been different economic, political, social, and cultural interests, values and ideologies in society, there have been debates and fights over the access to positions and opportunities and over the role and function played by assessment in that context.

This leads us to conclude that: assessment in mathematics education exists, is here to stay, serves a variety of purposes (not necessarily compatible), and is closely connected to controversy.

The *purposes* of assessing students' performance in a given subject, especially in mathematics, seem to fall into three different but not independent main categories: the provision of *information*, the establishing of bases for *decisions or actions*, and finally, the shaping of *social reality*.

Provision of Information

The fundamental purpose of assessment in mathematics education is to provide information to some constituency about matters related to mathematics education and its outcomes. The ultimate end — which may be far away — is that the information demanded and supplied should help the constituency in making certain decisions and taking subsequent actions, even perhaps the decision not to do anything at all.

The information sought and communicated to recipients is a function of several variables including: the content of the information gathered; the ways in which it is gathered, recorded, processed and communicated; the people or the system gathering it; the intended recipient(s) of the information and their needs. This section deals with the purpose of providing information to someone, and concentrates on information as a function of the recipient(s).

Assessment of students' mathematical performance (achievements and capabilities) may serve to provide information to different types of recipients.

Information to *the individual student*

- About his or her performance in *absolute terms*, i.e. *with respect to the given course or curriculum*, in view of its inherent requirements and challenges, either during the course or at the end of it. This may include information on the student's strong and weak points as regards his or her mastering of content; qualitative and quantitative level of activity; working habits and so forth. The goal may be to give either a *static/accumulative* picture of the student's performance (what does he or she master *now?*), or a *dynamic/differential* one (what are the changes in what the student masters *now* compared to *before*?).

- About his or her performance *in relative terms*, i.e. *relative to other students* in parallel situations, e.g. in the same group, class, school, district, country, region, or in the world. Again the picture sought may be static or dynamic.

The ultimate end of supplying students with this sort of information is to assist them in further improving or developing their situations, in absolute or relative terms, by equipping them with instruments for monitoring and controlling their own work and learning and the results thereof. Assessment conducted with this purpose is often called *formative assessment*.

Information to *the teacher*

- About the situation and the development, in absolute or in relative terms, of the *individual student*. This may serve several different aims: to enable the teacher to inform and advise the student; to assist the teacher in assessing his or her teaching and its outcomes in relation to that particular student, in order to adjust, develop, or fundamentally change it to what would better meet the needs of the student; to take decisions and actions to influence the student's behavior; to provide a basis for the teacher's reporting to the parents, the school, the authorities, further educational institutions, and employers of the performance of the individual student.

- About the outcomes of his or her teaching in relation to *groups of students* (maybe all of them, or maybe only those belonging to a certain category, class, or age level), in order to judge whether it should be changed or developed in certain directions.

Information to *the system* (An aggregate term, it is used here to encompass the parents, the school as an institution, recipients of school leavers (further educational institutions, employers), administrative and political authorities.)

- About the performance of the individual student to provide a basis for taking (or not taking) decisions or actions concerning *selection and placement*, e.g. promotion, recruitment to further education, licence to exercise a profession, employment with respect to that student (if this sort of information is supplied through a terse, integrated summary of the student's performance, the assessment is often called *summative*);

- About the distribution of students' performance as an input to the *evaluation* of teachers, institutions, districts, programs, curricula. (However, as stated earlier this is not in focus of this book).

Taking Decisions and Actions

In any society a multitude of decisions and actions are taken to *filter* and *select* individuals for opportunities, positions, jobs and privileges.

The term "society" is to be taken, here, in its widest meaning, to include the individual as a private person, as a citizen, or as a decision maker in the public or the private sector, as well as the political-economic system at large, encompassing structural, organisational, and cultural traditions. The term system as defined above, thus, designates a proper subset of society. The bases for, and the ways in which decisions on selections are made, are not simply a neutral matter of practicalities but are deeply rooted in the spectrum of material conditions, ideas, values, and traditions that are characteristic of a given society.

Sometimes selection decisions are taken and carried out entirely informally from a basis of personal acquaintance or knowledge, spontaneous impressions, sheer coincidence, position in the power or influence hierarchy. Sometimes more formalized bases are established which involve assessment procedures and results. In most societies both informal and formal selection modes are widely used, but not in the same context. Why does society engage in applying formal, assessment-based modes of selection? A balanced answer to this question seems to comprise several components.

In some professions, subject-specific competence really matters. A few obvious examples are the teaching of mathematics, research in mathematics, the exercise of the application of mathematics and mathematical modeling in special contexts. In such cases it makes sense — in principle, if not necessarily in practice — to operate with the notion of *minimum competence*. It may well be very difficult to agree on a definition, let alone a concrete representation, of that notion, for instance because it varies with time and space. However, there will probably be universal agreement that it would be ill advised to employ someone possessing no mathematical knowledge whatsoever in any of the three types of jobs just mentioned. So, it is not unreasonable to imagine the existence of a minimum competence related to certain professions.

This leads us to the notion of a *licence* (or *certification*): In any society, the officially acknowledged exercise of certain professions, e.g. that of medical doctor, mathematics teacher, or actuary, requires some demonstration of minimum competence. Due to the complexity and the division of labor characteristic of modern society, such competence is provided by educational institutions rather than by the agencies, companies, or institutions, that employ such professionals. These, in turn, therefore need to be equipped with some form of a "passed" stamp or a licence to practice their profession. The decision of which candidates should or should not be awarded a passed stamp inevitably implies the application of some form of assessment procedure.

Such binary decisions (passed/non-passed) are not only used to determine who can be licensed to practice professions. In most societies they are also crucial factors in determining who can be promoted to the next stage of opportunity or career, e.g. to the next class in school, to a selective educational stream or institution, to the next rank in the hierarchy, and so forth.

In addition to needing tools to help determine who should be admitted to take part in a given selection game, segments of society often want to choose between candidates who fulfil the minimum requirements for entering the game. This leads to the idea of *ordering* and, especially, *grading* candidates. Ordinarily the order relation (which may be linear, in which case it is often called *ranking*, but does not have to be) is established on the basis of performance, or at least with performance as an essential component. Again, this requires that some form of assessment procedures are devised and activated. The main purpose of ordering candidates is to provide a flexible instrument for selection, in particular in contexts where privileges are scarce, e.g. jobs and salaries, places in schools and universities, grants and awards, free accommodation, and membership of clubs, societies, or committees.

Now, normally society will need to have considerable confidence in the assessment procedures applied to reach a passed/non-passed verdict. In particular, society needs to be confident that someone equipped with a "pass" really satisfies the conditions of minimum competence; otherwise these would not be meaningful. Societies which can afford it will tend to worry more about granting a licence to someone who does not meet the standards set by society than about denying it to someone who actually does meet them. In any society there will be a serious interest in striking the right balance between passed and non-passed, in other words in devising assessment procedures which are valid and reliable to that end.

The situation seems to be somewhat different when it comes to the ordering of performance. Of course, valid orderings of performance, if they existed, would be highly valued as important social instruments to assist with the selection of the best in a group of able candidates. However, even if there is serious doubt and general skepticism concerning the ordering schemes in current usage, in particular regarding their validity, society tends, nevertheless, to use them frequently without worrying too much about their drawbacks. Four rationales for this are possible.

(1) If we do not possess completely satisfactory assessment modes that meet all our needs we have to content ourselves with as good an approximation as we can find, hoping that it is better than nothing. Here "nothing" means inexplicit and informal modes of assessment, not the absence of assessment, since judging of performance will always be carried out.

(2) Regardless of all well-justified reservations we might have towards the validity of current explicit and formal assessment modes, these *do* have at least one advantage over the inexplicit and informal ones which would otherwise be applied. By being explicit and formal, they become visible and hence subject to analysis, debate, and perhaps change. In that way, they contribute to diminishing the anarchy of arbitrariness, the reign of tacit criteria and rules, and even the nepotism that can result from implicit assessment.

(3) If the rules of the assessment game are blurred, those who can interpret the rules are better off than those who cannot. This factor tends to favour members of certain segments of society — defined by economic position and capacity, rank, gender, family — more than others. It, in turn, contributes to furthering an "unequal opportunities society". Therefore societies holding or following egalitarian values have often utilized formal assessment modes for selection to certain types of positions and jobs.

(4) Even if those engaged in taking decisions or actions based on formal procedures of assessment do not have confidence in the validity or overall quality of the procedures applied, society might still want to use these procedures as a tool for legitimizing decisions or actions toward criticism. Whether the assessment modes have actually been applied to reach the decisions or actions in question, or they are just used to cover up decisions made on quite different grounds, is not important. The essential thing is the legitimization invoked by referring to decisions and actions as having resulted from the use of objective assessment procedures.

It is not surprising that students' mathematical performance is an object of crucial interest in assessment utilized to promote and select (some of) them for further mathematical activity or for mathematical professions. What is a remarkable fact, however — unique to mathematics, so it seems — is that in many countries the assessment of mathematical performance plays an essential role in promoting, filtering and selecting students for studies which have no mathematical content at all. As Jean-Pierre Kahane once put it in a conversation, "The problem, perhaps, is not so much assessment in mathematics as mathematics in assessment."

This use of assessment in mathematics for non-mathematical purposes seems to rely on two assumptions that are not mutually exclusive. I use the word assumptions because this is not the place to discuss their theoretical or empirical justification. Whether they should be considered as observations, beliefs, facts or scientific results is not an issue here.

The first assumption is rooted in classical faculty *psychology*, which sees mathematical capability as closely connected to general cognitive and intellectual capacities and skills relevant and significant in the pursuit of any analytic intellectual activity. It does not matter if mathematical capability constitutes a cause or an effect of general intellectual capacity, as long as the former can serve as an indicator of the latter. The second assumption is more *educational* in nature. It emphasises that someone who has proved succesful in difficult mathematical studies has demonstrated a level of ability, commitment, self-discipline, and perseverance that will increase the probability of success in any study, mathematical or not. Whether success can or cannot be attributed to a special mathematical talent, if such a thing exists, or to hard work is less important. What matters is success in a difficult and demanding discipline.

Once again, although we have chosen to limit the scope of this book to dealing with assessment rather than with evaluation, it should not pass unnoted, that in most societies, one of the important purposes of assessment is to monitor and control the educational system in general, and the institutions, the teachers and their teaching, as well as the various curricula, in particular. Thus, in many places students' assessment results are used as a factor in decisions concerning teachers' careers, promotion, hiring and firing, salary adjustment. They are further used to provide a background for modifying or reforming the curriculum — or to keep it unchanged.

Shaping of Social Reality

Naturally, when assessment in mathematics education is conducted in accordance with the purposes outlined above, a strong influence is exerted on the social reality of students, teachers, parents, schools etc. The actual modes and procedures of assessment contribute to shaping the conditions and lives of a lot of people. But, many of these consequences are side effects which are not directly intended as an explicit aim of assessment.

However, there is reason to believe that society also intends to give assessment the task of shaping the larger social reality by disciplining students, teachers, institutions; creating a stick-and-carrot climate of commitment to hard work; competition and struggle; acceptance of — or at least adoption to — divide-and-rule reality; subordination to power, authority, and ideology. The degree to which such intentions are emphasized and the subtlety with which they are pursued, vary with the society, but any society holds it and has to hold it, to some extent, in order to have its citizens subject themselves to the often difficult, stressing and frustrating conditions inherent in the competition for education, job, position, and material wealth, in one word, for success.

Summing Up

The considerations offered above may be summed up in the following *purpose matrix* (Figure 1) displaying the relationsship between the purposes of assessment (the rows) and the target groups aimed at by those purposes (the columns). The cells of the matrix represent how the corresponding purpose concerns the corresponding target group. Entries in parentheses refer to matters that have to do with evaluation rather than assessment proper. Arrows should be read as "feeds" or "provides a basis for".

3. ASSESSMENT MODES IN MATHEMATICS EDUCATION

We shall use the expression **assessment mode** as a compressed term to indicate the full spectrum of components in an assessment model that is or could be implemented. Any specific assessment mode installed in a given educational system somehow reflects, although not necessarily in an isomorphic way, the purposes attached to assessment within that system, as discussed in the preceding section and depicted in the purpose matrix.

Metaforically speaking, a given assessment mode may be thought of as a "vector" with a set of components. A partial list of these include the following:

- The **subject** of assessment. *Who is assessed*, i.e. On which *unit* of students is the specific assessment under consideration carried out, the individual student, a working group of students, the class, the cohort, the institution etc.?

- The **objects** of assessment. *What is assessed*, i.e. Which types of mathematical *content* (including methods, internal and external relations) and which types of student *ability* to deal with that content form the objects of the assessment at issue.

recipient/ subject/ agent purpose conc.	the student (St)	the teacher (T)	the system (Sy)
information to St, T, Sy, resp., about:	his/her own performance	his/her students' performance * his/her own teaching	the individual students' performance * (student populations, teachers, institutions, curricula)
decisions and actions taken by St, T, Sy, resp., concerning:	controlling his/her own learning activity	controlling his/her teaching * advising and controlling the individual student's activity and placement	(monitoring and controlling of teachers, institutions, and curricula) * filtering, selecting and placing the student
shaping social reality for St, T, Sy, resp., as regards:	priorities of investment of effort and time * working habits and attitudes * attitudes towards competetion and social order	choice of content, material and methods of teaching * classroom environment * priorities of investment of effort and time * working habits and employee attitudes * attitudes towards competetion and social order	maintenance and self-preservation of the system * controlled development of the system

Figure 1 *Purpose matrix*

- The **items** of assessment. *What kinds of output are assessed*, i.e. How are the objects developed into assessment tasks and demands? What specific items is the assessment subject supposed to produce as output to be assessed?

- The **occasions** of assessment. *When does the assessment take place*, i.e. **at what** instances during, in, or after the course, term, program, or entire study is assessment information gathered?

- The **procedures and circumstances** of assessment. *What happens, and who is expected to do what* on the assessment occasions?

- The **judging and recording** in assessment. *What is emphasized*, what are the criteria and procedures for judging the assessment items, and *what is recorded* to form the results of the assessment process?

- The **reporting** of assessment outcomes. *What is reported to whom* about the assessment process and and its results, and in what shape and format is the reporting carried out?

In the paragraphs to follow we shall elaborate on each of these components.

The Subject of Assessment: Who is Assessed?

According to the definition of assessment suggested for this study, a first answer is simple: the individual student is the *ultimate subject* of assessment in mathematics education. In addition to being almost a matter of definition and formality this answer is also a consequence of the various purposes of assessment. So, in the majority of cases, the individual student is not only the ultimate subject of assessment but is also the *immediate unit* on which assessment is conducted.

In recent decades, however, group work has been introduced in mathematics education in quite a few places around the world. This has implied that the immediate subject of assessment is now sometimes a working group of students. Still, this does usually not mean that in such cases the individual student has been replaced by the group as the ultimate subject of assessment. Often the working group simply serves as a narrower and more comfortable setting, "a wrapping", within which assessment of the individual student may then be carried out on the basis of the group work. This may happen either indirectly, by restriction to the individual student of an overall judgement of the group work as a whole, or directly by assessing the individual student in relation to his or her contribution to or mastery of the group work. We seldom encounter examples in which the working group as an aggregrate unit remains the ultimate subject of assessment without being broken down to the level of the individual.

When larger collectives of students, classes, schools, age cohorts, are the subjects of assessment in mathematics education it is often because some form of competition between the collectives is involved, or because the focus of interest has in fact been shifted from the assessment of students to the evaluation of schools, curricula, teachers, or systems.

The Objects of Assessment: What is assessed?

Of course, this varies greatly with the place, the teaching level, and the curriculum at issue. Viewed internationally, the predominant content objects assessed seem to be the following:

a) *Mathematical facts* i.e. definitions, theorems, formulae, certain specific proofs, historical and biographical data (encountered only on very rare occasions, though).

b) *Standard methods and techniques* for obtaining mathematical results (e.g. qualitative or quantitative conclusions, solutions to problems, display of results).

c) *Standard applications* i.e. characteristic types of extra-mathematical situations, usually closed, treated by collections of well-defined mathematical tools.

In an increasing but limited number of cases, objects of assessment include

d) *Heuristics and methods of proof* i.e. ways of generating mathematical results in non-routine contexts.

e) *Problem solving* of non-typical, perhaps open-ended, perhaps complex problems, and

f) *Modeling* of open, complex, ill-defined, sometimes real and even authentic, extra-mathematical situations belonging to other subjects, or practice areas, treated by whatever mathematical means at one's disposal.

We rarely encounter

g) *Exploration and hypothesis generation*

as objects of assessment.

With regard to the assessment of types of student abilities, the focus is on types of abilities that match the various types of mathematical substance just mentioned. What is most frequently assessed then is students':

- *Knowledge of facts*, with particular attention being paid to accurate recollection of the facts; their ability to quickly, precisely and coherently express these facts; and their ability to adequately select and combine facts in various contexts;

- *Mastery of standard methods and techniques* to obtaining mathematical results, mostly in standardized contexts, with particular attention being paid to the range of methods and techniques students can activate, combine and flexibly employ, and to the range and complexity of contexts they can successfully deal with;

- *Performance of standard applications* of mathematics in typical situations, again with particular attention being paid to the range and complexity of the situations they can handle, and to the originality and depth with which they can treat these situations.

Naturally, also the less usual content object has its counterpart in a type of student ability to be assessed, e.g. students' ability to activate or even create heuristical procedures and methods of proof; to solve open, complex problems; to perform mathematical modeling of open, extra-mathematical situations; and to explore situations and generate hypotheses.

What could be the philosophy behind the fact that this set of content and ability items have been selected as objects of assessment in mathematics education?

First, all of the items outlined above *are* widely considered to be essential representations of what mathematics and mathematical activity are really about. Items a)-g) and the corresponding abilities *do* embody characterstic aspects of mathematics and the exercise of it, whether described in process or in product terms, without claiming to be exhaustive. The first items in the list, a)-c), may be seen as emphasising rather unsophisticated and routine features of mathematical work whereas the remaining ones are more sophisticated and demanding. Nevertheless, this does not prevent the former from being necessary, even if not sufficient, indications of the presence of mathematical knowledge, insight and capability.

If items a)-g) are all so important why, then, are current assessment modes in mathematics education so often restricted to dealing only with the first few, i.e. a), b) and c)? One possible answer to that question could be the following:

First, as we have just seen, a), b), and c) are fundamental instances of mathematical knowledge, insight and capability. It is therefore only natural that the former are made objects of assessment as instances of the latter. Second, we think we possess means for assessing items a), b), and c), while at the same time observing a number of traditional requirements such as (see Niss, in press): Assessment should be *reliable*, i.e. there should be no room for different assessors to seriously disagree on the judgement of a product or process performed by a given student. Assessment should be *just*, i.e. students in similar situations should be assessed on equal grounds, e.g. under uniform test conditions with time constraints. Assessment modes should be implementable on a *large scale*, i.e. make use of standardized schemes, in order to treat large populations of students in the same way. Assessment modes should allow for easy and objective *marking, grading, and comparison* of student performance. Assessment should be *inexpensive*, i.e. assessors should spend only a limited amount of time on assessing each student.

In contrast, it has proved far more difficult to devise tools for assessing items d)-g) that lie within the boundary conditions set by the traditional requirements listed above. Although there is general agreement that the inclusion of these items as objects of assessment would bring new dimensions of *validity* into the assessment of mathematics, many argue that we cannot just abandon the traditional requirements because doing so would introduce an element of unreliability or arbitrariness that could result in undesired consequences. Others argue that if we assess only items a), b), and c) and continue to leave items d)-g) outside the scope of assessment, we not only restrict ourselves to assessing a limited set of aspects of mathematics, but also contribute to actually creating a distorted and wrong impression of what mathematics really is. This line of reasoning has led people in the latter quarter to insist that we should seek new ways to assess these items and be prepared to abandon some traditional requirements for the benefit of greater scope and validity in assessment.

The Items of Assessment: What Kinds of Output are Assessed?

It has been a dominant characteristic of mathematics teaching and learning that *the task* forms the center-piece of attention and activity. Stieg Mellin-Olsen (1990) has found that, to a large extent, mathematics education focuses strongly on what he calls "the task discourse". What do we mean by "a task"? A task may be defined as an *orientated activity*, i.e. a set of actions orientated towards undertaking certain *missions* such as orders, proposals, or challenges.

A task is formulated orally or in a written form by someone (ranging from oneself to an anonymous subject), often by using the imperative mode: compute..., determine..., solve..., draw..., construct..., prove..., investigate..., explore..., describe... Tasks may also be formulated by means of questions: "Which of the following statements are true...?", "How many...?", "What if...?", "Is it possible that...?", "Does there exist...?", "Under what conditions...?", "What is the relation between...?" A task may be called *artificial* if it is set by someone who already knows how to accomplish it, and *authentic* if this is not the case. The goal of a task is simply to accomplish the given mission, whether artificial or authentic, and whether formulated in the imperative or in the interrogative mode.

The traditional categories of tasks are *questionnaires*, *exercises* and *problems*. All of these are defined relative to particular groups of addressees. A questionnaire contains a collection of questions that usually concern facts (such as definitions, properties, formulae, or results of computations) with one or a very small number of possibilities for answers to each question. Exercises involve primarily routine type considerations or operations in straightforward combinations, usually within the context of a well-defined curriculum. In contrast, problems involve considerations or operations, or combinations of such, which are non-routine relative to a given curricular context. Exercises and problems are not absolute concepts. What, to one person, appears to be a problem may be an exercise to another person (or the same person at a later stage of development).

In the context of teaching and learning mathematics, problems — or to be more precise, *problem situations* — may be of different kinds. Some problems are *closed* in that specific well-defined items and issues are to be tackled, and definite results subject to clear-cut requirements are to be obtained by utilizing tools which are also rather well-defined in the context, for instance by referring to certain segments of the curriculum, or even to chapters in textbooks. With other problems, the items and issues to be tackled are still well-defined whereas neither the results to be obtained nor the means to obtain them have to be limited to one or a few fixed types. Such problems may be called *open-response* problems. Finally, there are problems in which the problem situation is broad, complex, contains a multitude of specific (sub)problems, or is formulated in general — or even vague — terms. For such problems to be dealt with in a meaningful way, the first step is to interpret, specificy and make the problem precise. If this step is left to the individual(s) working with the problem, we may speak of an *open-ended* problem. Naturally, with open-ended problems it is also part of the task to identify the requirements that the results sought should meet and to, accordingly, choose the methods and techniques which are to be used to obtain them.

The reason that the notion of task occupies such a dominant position in the teaching and learning of mathematics is that it represents characteristic features of mathematical work and thinking in concrete terms, through organised and orientated activities. This is not only true for educational contexts but also, and even more so, for mathematics as a research subject. To sum up, we may perceive the task as an organisational vehicle for the exercise of mathematical activity, in the same way as the content and ability components discussed in the section *The Objects of Assessment* correspond to important content categories of mathematical work.

All this makes it easy to understand why tasks — and especially questionnaires, exercises, and problems — constitute the predominant *item of assessment* in mathematics education. They occur in various places in the curriculum: continuously, as part of the teaching and learning of mathematics in the classroom; in tests set locally by the teacher or the school on definite occasions during the term to monitor and control learning and teaching; and at examinations set locally or by central curriculum authorities to assess students at the end of a given stage in the educational system.

The ultimate purpose of all assessment tasks is the revelation of the level of knowledge, mastery and capability of the individual student in relation to the goals, aims, and requirements of the underlying curriculum. Very often the tasks are designed to reveal which facts (as defined above) the student knows (or does not know), which standard techniques he or she has mastered, and the range of situations to which he or she can apply these techniques. Particular emphasis is placed on the rapid and correct performance of computations and calculations based on formulae, either in a purely mathematical context or in a context involving standard applications of mathematics. Much less frequently, and mainly at higher or more specialized educational levels, tasks are designed with the purpose of revealing students' general mathematical ability, such as: their understanding of the network of relationships between concepts; their ability to combine flexibly concepts and results to deal with non-standard situations; their mastery of general, i.e. context-free, methods such as induction and indirect proof; their ability to analyze creatively situations which are new to them and to devise ways of tackling them; and their ability to generate, verify or refute mathematical hypotheses.

Even if most mathematics educators would agree that it would be valuable to design assessment tasks aiming at general mathematical ability also for ordinary pupils and students, this does not happen very often. There seem to be several reasons for this. First, it still remains a widely held assumption that general mathematical ability, as briefly outlined above, can be found or cultivated with only a few particularly apt students, so there is no point in assessing its presence with ordinary students by including it in assessment tasks. Second, the more complex abilities a task encompasses, the more difficult it is to interpret its outcome in a reliable way. This problem is aggravated if tasks are for large student populations. Third, irrespective of our wish to design tasks aimed at assessing complex mathematical abilities, students will still have to know and master the less complex facts and techniques, so assessing these, as a minimum, seems to be justified.

Mathematics educators who subscribe to a behaviorist standpoint sometimes claim that we shall never be able to uncover understanding or other highly complex abilities. What we can hope to do is to identify certain things that we want students to be able to *do* in well-defined categories of behavior and then devise tasks that allow us to measure to what extent an individual student can actually do them. According to this position, an assessment task should be conceived as an experimental probe inserted into a complex world, the nature of which is, in principle, inaccessible. In order to come to know something about this world, we have to act as though performing experiments in the natural sciences. One of the crucial requirements in this domain is that experimental probes have to yield unambiguous data so that the interpretation of the experiment does not rely on individual data but on patterns they constitute. Although in recent years behaviorist positions are held less strongly by mathematics educators, the construction of assessment tasks is still influenced by behaviorist philosophy.

One important consequence of the fact that the task constitutes the predominant items of assessment is that the tasks used point out (to students as well as to teachers) what the essential components of mathematics and mathematical ability are considered to be. That which is not represented in assessment tends to be perceived by those involved as less important. This might lead to an eventual withering of these components. Furthermore, what is used in tasks is frequently preserved in students' memories more easily than that which occurs only now and then, or never. In that way, assessment tasks filter and mould the perception of mathematics as a subject. This is just a special instance of the well-known general backwash effect of assessment modes on mathematics education as a whole.

In the majority of the world's countries, assessment tasks consisting of questionnaires, exercises, or problems are written ones both in the formulation and, more importantly, in the responses they require. This includes homework assignments, tests, and exams. So, students are supposed to produce output in the form of sheets of paper containing their written responses to the tasks set, whether filled-in questionnaire sheets, mathematical operations accompanied by explanatory remarks, or a proper text that exposes and explicates the task and its accomplishment. In these cases, the sheets of paper submitted by the student constitute the *output items* on which that student will eventually be assessed.

A written format, however, is not an essential feature of the tasks considered so far. In some places, tasks are sometimes set as oral tasks and the student is asked to accomplish them in front of the class, or just in front of the teacher or a group of examiners, and so forth. In contrast to written tasks, this oral format allows other persons present, e.g. fellow students, teachers or examiners, to put questions of clarification or examination to the student and to interview him or her. In some cases, however, the student is asked to deliver a full oral presentation without being interrupted or interviewed. With oral tasks, the output item to be assessed is the student's performance in its entirety, including his or her responses to questions, if these apply.

Even if questionnaire, exercise and problem tasks *do* dominate the world assessment scene there *are* other items on it. As a matter of fact, there is a growing demand for and development of new types of assessment items. The new types include: real academic *theses or dissertations*, genuine or simulated *articles* for journals or newspapers; *essays* on topics within or related to mathematics; written *project reports* in which the outcomes of thematic or problem-oriented investigations are presented; *portfolios* i.e. collections of various sorts of written material produced and kept by the student in relation to and during a given teaching/learning unit; *exhibitions* consisting of posters or physical objects, but also films, audio or video programmes, slide shows etc., or short specimens of *communication technology* (including computer soft-ware) that display and explain topics, themes or findings of investigations, or demonstrate phenomena and their interrelations.

A characteristic to most of these items is that they cannot be accomplished within the time constraints usually set for the traditional assessment tasks. They need a longer span of time to accomplish, often several months. It is natural for them to be incorporated into, or at least to run along with, everyday course activities: they become *coursework items* of assessment. They may also involve extra-institutional activities.

The Occasions of Assessment: When Does Assessment Take Place?

As assessment serves a variety of purposes the occasions on which assessment may take place vary accordingly. To begin with, we may distinguish between *continuous* assessment and *discrete* assessment using the category of time.

Continuous assessment takes place concurrently with, and is often integrated into, the teaching/learning unit at issue, and is based on the individual student's general performance and achievements, level of activity, working modes and so forth during the unit. So, in accordance with its nature, continuous assessment relies on coursework (whereas the converse does not need to be the case). That assessment is continuous does not imply that it takes place incessantly, only that no specific *assessment events*, each concentrated within a limited span of time, are involved in it. In principle, *continuous assessment* may serve all the purposes in the purpose matrix of the previous section. In practice, however, in most places its primary purpose is to inform students (and their parents) about their performance so as to help them control and adjust their learning activity. An almost equally important purpose is to inform the teacher about the outcome of his or her teaching in general — in order to adjust it if desirable — and specifically in relation to the individual student in order to advise and influence his/her actual or potential association with mathematics.

In contradistinction to what is the case with continuous assessment, *discrete assessment* is conducted on the basis of specific assessment events. Such events may occur at the beginning or during a course. The items assessed at such course-internal events may be coursework items, for example, portfolios or (selected) homework assignments, or specially designed *(local) tests*. Assessment events also occur at the end of a course, term, year, or of a major stage in the educational system (end of primary, lower, or upper secondary school, completion of a degree at a tertiary institution); we shall refer to them as *examinations*. Examinations held at the end of a major educational stage are often called *final examinations*. A special type of discrete assessment is the examinations held in some countries to select students for admission to certain educational institutions. Not surprisingly, such examinations are known under the name of *entrance examinations*.

As is the case with continuous assessment, discrete assessment may also serve all the purposes in the purpose matrix. There is, however, often more emphasis on providing information and assistance to the system than to the individual student even if this latter purpose is still in function. Tests and non-final examinations serve, in particular, to inform the teacher, the institution, and the parents, and to assist them in making decisions on pass/fail, or placement, or to influence the behavior of the student, within the given institution or segment of the educational system. Final examinations mainly serve to permit conclusions as to whether the individual student has succeeded or failed in completing a certain educational stage and to provide information (possibly supplemented with ranking or grading information if available) to other educational institutions, employers, and agencies in society. This information is likely to be instrumental when decisions on selection, placement and employment of the student are to be made.

It should be noted that continuous and discrete assessment, although disjoint by definition, are not mutually exclusive. They may well, and often do, co-exist. It is not unusual for a teacher to conduct both continuous and discrete assessment of the performance of each student in a class.

The Procedures and Circumstances of Assessment: What Happens Where and Who is Expected to Do What?

As quite a few of the procedures and circumstances of assessment, especially regarding their details, are matters of practicality and local tradition rather than of principle, they are subject to considerable variation with country, curriculum, and institution. As far as *continuous assesment* is concerned, students produce the items of assessment, whether material (e.g. written texts) or immaterial (e.g. activity level in class) either directly in the classroom or as homework. The judging of these items, as well as the recording and reporting of the judgement in whatever relevant format, normally are carried out by the individual teacher without the involvement of anybody from outside.

Tests belong to the category of discrete assessment and are associated with assesment events. The items involved may, again, be both material (e.g. written examination papers) and immaterial (e.g. an oral interview). Tests may be conducted by the individual teacher within the setting of the classroom, or by groups of teachers, or by the institution.

Typically, tests are conducted under standardized conditions. The individual student is expected to work alone under time constraints with a given task, set locally by the student's teacher or by teachers at the same institution, and to have delivered a — typically written or oral — product by deadline. Ordinarily, the time scale of the test ranges from a fraction of an hour to a few hours. Such tests, whether written or oral, take place in the student's institution, and often the student is prevented from consulting textbooks or other people during the test.

Even if this depicts the typical test situation, a wide spectrum of variations and modifications exist around the world. Sometimes students may bring their textbooks with them to a written test. Sometimes the student is given one or more days to accomplish the task as a take-home-test in which case the student is not, in practice, prevented from seeking advice from somebody else. Sometimes students are allowed to accomplish their tasks in teams. Sometimes students are given an opportunity to prepare themselves on a specific task which is to be finished orally or in writing under test conditions that in other respects are the standard ones.

Again, as was the case with continuous assessment, the judging, recording, and reporting of items produced in tests are normally carried out by the individual teacher, perhaps assisted by colleagues at the same institution acting as moderators as far as the judging is concerned.

When it comes to *examinations*, final or non-final, the formats used are typically the same as for tests, except that in many places *external professionals* — i.e. teachers from other institutions, representatives of external curriculum authorities, for instance ministries or school boards, independent assessment agencies, and so forth — are involved in some way or another. External professionals may be in charge of or in other ways involved in setting the examination tasks, in judging the individual student's performance, and in moderating and homogenizing judgement criteria, or recording/reporting schemes.

In examinations, the standard test conditions are even more dominant than is encountered with tests. In both cases the standard test conditions are conceived as means of paying respect to the traditional requirements listed in a previous paragraph: Assessment conditions should be just and allow for reliable judgement as well as for objective and easy marking, grading, and comparison of students' performance.

The Judging and Recording in Assessment: What is Emphasized

What is emphasized as far as judging and recording of performance are concerned? Once again, the dominant feature is *variation* with circumstances and traditions, but certain features, including the following, do seem to manifest themselves as common to a great deal of assessment schemes. At the most general level, the majority of mathematics educators involved in assessment would maintain that what should be emphasized when *judging* mathematical performance is the extent to which essential mathematical virtues — knowledge, capability, and aptitude, as these are reflected in the given curricular context — are present with the students who are being assessed. More specifically, what is emphasized is students' mastery of the categories of content and ability explicitly or implicitly established in connection with the curriculum, i.e. content types a)-g) and the corresponding types of ability as outlined in the paragraph *The Objects of Assessment*.

It is, however, difficult to pursue and maintain this emphasis in a general and extensive sense. Instead, the content and ability types occur in specific forms, filtered through, modified, and shaped by the concrete assessment items used, as well as by the entire set of assessment circumstances. What happens in practice, therefore, is that judgements are concentrated on a restricted collection of mathematical virtues, namely those immediately relevant for dealing with the assessment items. So, the narrower the spectrum of content and ability categories that are represented in the assessment items, the narrower the range of judgement emphases.

This is particularly pertinent when *tests* and *examinations* are conducted under usual test conditions with time constraints, and even more so when usual written examination papers are involved. In such contexts the dominant emphasis is on *correctness*, and often on the *degree of correctness*. The latter concept presupposes the notion of *completeness* and *absolute correctness*, i.e. that it makes sense to speak of a task or a set of tasks as being accomplished, or, equivalently, of a final, complete and absolutely correct result as being attained. On that basis, a given performance is considered to represent some fraction of complete correctness, which leads to registering the outcome as some degree (possibly given in quantitative terms such as percentage) of an absolute unit. Tasks not completed are taken to display a corresponding lack of correctness.

If a task is judged to be finished to completion and to absolute correctness this is often assumed to imply that the student has mastered all the underlying knowledge, insights, skills, and methods that are considered necessary to reach this point of correctness. Similarly, partial correctness is taken to reflect a corresponding partiality in the student's mastery.

The emphasis on completeness and correctness entails a derived emphasis on students' mastering of the *methods*, *techniques*, and *procedures* involved in arriving at the goals.

Roughly speaking, the strength of these emphases tends to assign secondary importance to other matters. *Reasoning, explanation, exposition*, and *communication* are attributed importance mainly to the extent that these factors assist or hamper, respectively, assessors' gauging of degrees of correctness. It also happens, though less often, that these factors are emphasized as being valuable in their own right.

Within a given curricular context certain approaches are considered more acceptable than others. Non-standard approaches and hence forms of creativity, inventiveness, and originality which are less transparent at a first glance and do not immediately manifest themselves as lying within the established norms, are less favored. In effect, assessment often seems to emphasize *normality*.

As a general consequence of the emphasis on completeness and correctness, the single most important factor in determining the outcome and judgement of assessment conducted under test and exam conditions is *speed*, or to be more precise, *efficiency*. If a student is capable of accomplishing a set of assessment tasks but not within the given time constraints, he or she will be judged as less capable, not infrequently as much less capable, than would be the case under different circumstances. This is a widely recognized fact that generates a lot of disagreement. To some educators, this fact implies that the assessment modes involved suffer from a serious lack of validity, and are hence misleading. Moreover, these modes are likely discourage students who are mathematically able but work slowly from undertaking the additional mathematical studies they could well manage. Others maintain that even if speed is not perhaps an independent goal, it *is* an important indicator of mathematical aptitude, and that if two students are capable of accomplishing the same tasks, but at different speeds, it is fair to consider the quicker a better student than the slower.

In accordance with the emphasis on completeness, correctness and efficiency, it is usually the case that what is *recorded* as a the outcome of assessment is an aggregated result, not infrequently a *single element in a one-dimensional marking scale*. The scale consists of symbols and may be quantitative (i.e. involve numerical properties of numbers and not just their capacity for being symbols) or qualitative. A 0–1 (pass/non-pass) scale forms the one extreme, a percentage scale the other, with a variety of multistep scales in between. Sometimes what is recorded is a "low-dimensional vector" of one-dimensional results, for instance one result for coursework and one for a test or exam, or one for a written test and one for an oral performance; or one for factual knowledge, one for procedural skill, and one for problem solving. When non-traditional abilities, such as open-ended problem solving or modeling, are the objects of assessment, there is a tendency in some places to operate with result vectors of higher dimensions corresponding to splitting a complex ability at issue into components or sub-abilities.

Even if aggregate results given in short-hand or symbolic form dominate the ways in which assessment outcomes are recorded, there are examples of more composite and comprehensive ways of recording assessment outcomes, for instance by statements that characterize students' performances in qualitative terms in an ordinary language. Such examples are much less frequent in the context of tests and examinations — in particular final ones — than with continuous assessment and coursework assessment. It still seems to be the case, moreover, that institutions, authorities, employers, and so forth, are reluctant to receive extensive information of assessment results. They prefer to obtain records that are as aggregated as possible.

Regardless of the number of dimensions used in recording the outcome of assessment, each component will be of one of two different types: *absolute* or *relative*. If the individual student's achievement is assessed according to pre-established criteria or standards determined solely in content or process categories and with no comparisons to other students' achievements, the assessment outcome will be recorded in absolute terms. Frequently, they will be labelled *criterion-referenced* assessment instead of absolute assessment. If the individual student's achievement is assessed in relation and comparison to the achievements of a larger population of peer students, the assessment outcome will be recorded in *relative* terms and sometimes called *norm-referenced*. Although the two ways of recording assessment results are distinct in principle, they are often interwoven in actual practice. So, it is not unusual that what is recorded for each of the dimensions is an ill-defined, perhaps unconscious, mixture of absolute and relative recordings.

When it comes to the physical aspect of the recording, assessment results are almost exclusively recorded in written form. Normally only the assessor(s)' final records of results are filed, whereas notes, comments, and minutes are not. In the case of written assessment items, papers produced by the individual student are sometimes preserved. It seems that student portfolios are seldom filed. Assessment events, including oral ones, are seldom preserved in audio- or video-tape form.

The Reporting of Assessment Outcomes: What is Reported to Whom?

What is reported to whom about the assessment process and its results? Insofar as assessment is conducted in order to provide information to the student, the teacher, and the system, to assist with decisions and actions, assessment outcomes are reported to these categories of recipients. Naturally, the central piece of what is reported is formed by what is recorded.

What is typically reported (ordinarily in writing) to those segments of the system that are not close to the student — e.g. authorities, employers, further educational institutions, and so forth — is the assessment record in its most aggregate and terse form (marks and grades, accompanied, perhaps, by general statements). One consequence of this is that decisions and actions of the segments of the system that are based on assessment results reported in a very terse form, are limited in scope and quality by the extent and quality of the information recorded and reported.

The assesment reports presented to system segments closer to the student (his or her educational institution, teachers, parents) will often contain more information than the assessment results recorded and filed. Such additional information can include nuances, explanations, comments, or suggestions which provide a broader and more comprehensive background for decisions and actions. It is passed on orally, in personal conversations or meetings, and its details are not normally recorded.

If the assessment at issue has been conducted by the student's own teacher — i.e. not by authorities, agencies, or other institutions, in which case the teacher will be part of the system — that teacher will be in charge of the reporting to the system and usually also to the student. When it comes to reporting the outcomes of assessment to the student assessed, it is often the case that the recorded results are revealed to the student with a variety of explanations, comments, and advice typically brought about in personal conversations with the student. The purpose of giving such additional information to students is to provide them with assistance in monitoring and controlling their learning activity and behavior, or to counsel them in choosing between different options of further education or career.

4. ABOUT THIS BOOK

As stated in the first section of the present chapter, an essential background for mounting a rather extended ICMI study on assessment is an increasing mismatch in recent years between the current state and development of mathematics education, on the one hand, and the predominant assessment philosophies, modes, and practices on the other. Assessment involves a multitude of deep and difficult theoretical and practical problems some of which have been outlined in the preceding sections. We shall conclude this chapter by distilling these problems into one *fundamental dilemma*:

> Premise: Assessment in mathematics education tends to have a strong impact on the objects assessed, often to such an extent that they become distorted or even destroyed.
>
> Premise: What is not assessed in education becomes invisible or unimportant.
>
> *Dilemma*: How can we assess the essential components of mathematical knowledge, insight, thinking, creativity, problem solving, and overall capability without seriously distorting them?

In the following chapters of this book a wide variety of manifestations of this dilemma are analyzed, and possible ways of overcoming them are discussed. Thus, it is the purpose of this study to explore and critically examine assessment in mathematics education and its effects, in other words to *assess assessment*. This is done by analyzing general notions, concepts and functions of assessment in mathematics education, by examining current prevalent assessment practices, by identifying key issues and challenges, and by suggesting directions for future development.

The first set of chapters deal with aspects of assessment in mathematics education in fairly general terms. *Jeremy Kilpatrick's* (USA) chapter gives a historical overview of assessment and its relations to educational and psychological notions and positions. *Geoffrey Howson's* (UK) chapter takes a historical approach as well but concentrates on the relationship between assessment and society. Also the chapter by *Jim Ridgway & Don Passey* (UK) focuses on the societal aspects of assessment with a particular emphasis on its socio-political functions in different types of societies, viewed from an international perspective. Some of these emphases are also present in *Peter Galbraith's* (Australia) chapter which, however, deals particularly with general epistemological ramifications, and related paradigms, for assessment. *David Wheeler* (Canada) investigates the implications for assessment of epistemological positions toward mathematics as a subject, whereas *Thomas Romberg* (USA) focuses on how mathematical knowledge and insight is gained and discusses what consequences this should have for assessment philosophies and practices. *Antoine Bodin* (France) attempts to answer the question "What does it mean to assess in mathematics education?" by giving a detailed analysis of students' responses to a variety of assessment items, while *Stieg Mellin-Olsen* (Norway) discusses how the student can become a subject in assessment rather than only an object as is ordinarily the case in both theory and practice.

The next two chapters, by *Herbert Ginsburg, Susan Jacobs & Luz Lopez* (USA), and *Bengt Johansson* (Sweden), respectively, come closer to the assessment of the individual student, in particular with respect to the diagnostic assessment of young students' arithmetical thinking.

In *John Izard's* (Australia) and *Malcolm Swan's* (UK) chapters the focus is on identifying and answering new challenges to improving the design and balance of mathematical assessment practice with particular regard to a multi-faceted mathematics curriculum.

A group of three chapters deal with large-scale performance surveys and comparisons. *Derek Foxman* (UK) presents main findings in the Assessment of Performance Unit's surveys 1978–87. *David Robitaille & Stuart Donn* (Canada) present the background, philosophy, and the design of the Third International Mathematics and Science Study, whereas *Gila Hanna* (Canada) examines aspects of validity connected to international performance comparisons with reference to items included in Second International Mathematics Study.

In the final chapter of this book *Norman Webb* (USA) outlines the requirements and main components of a theory of assessment of students' knowledge of mathematics.

Acknowledgements

A great many individuals have played important roles in creating this study. First of all Claudi Alsina (Universitat Politècnica de Catalunya, Barcelona, Spain) who was the protagonists in organizing the ICMI Study Conference on Assessment in Mathematics Education and Its Effects, and the members of the international Program Committee, many of whom gave very valuable advice in the early stages of the editorial process. It is important to underline that even if only a minority of the participants in the Study Conference have contributed to the two studies resulting from it, all participants' work has exerted an important influence on the editing, the format, and the content of both books.

The Executive Committee of ICMI wishes to express its sincere thanks to Unesco, the Division of Basic Sciences, for having sponsored the Study Conference with a most helpful grant.

Special thanks go to Thomas A. Romberg and Joan Daniels Pedro, the National Center for Research in Mathematical Sciences Education, University of Wisconsin-Madison, USA, for their assistance with the final polishing of the articles. In particular, the Editor is deeply grateful to Joan Daniels Pedro who did a magnificent job in polishing and harmonizing the English of the chapters and homogenizing their formats. However, any remaining linguistic or editorial flaws are solely the responsibility of the Editor.

Many thanks are further due to Kluwer Academic Publishers, in particular to Peter de Liefde and Nicola Berridge, who have been most cooperative and flexible in the whole process of preparing this book. Last, but not least, the Editor wants to thank his son Henning Niss who was in charge of producing the camera-ready manuscript.

NOTE

A modified version of Section 1 of this introduction is contained in the other ICMI Study on assessment in mathematics education, *Cases of Assessment in Mathematics Education: An ICMI Study* (Niss, 1992).

REFERENCES

Churchhouse, R.F. et al. (Eds.) :1986, *The Influence of Computers and Informatics on Mathematics and Its Teaching*, ICMI Study Series, Cambridge University Press, Cambridge

Commission Internationale de l'Enseignement Mathématique: 1990, 'Assessment in Mathematics Education and Its Effects: Discussion Document' *L'Enseignement mathématique*, 2e série, **36**, fasc. 1–2, Janvier-juin 1990, 197–206

Howson, A.G., Kahane, J.-P., Lauginie, P. & de Turckheim, E. (Eds.): 1988, *Mathematics as Service Subject*, ICMI Study Series, Cambridge University Press, Cambridge

Howson, A.G., & Kahane, J.-P. (Eds.): 1990, *Popularization of Mathematics*, ICMI Study Series, Cambridge University Press, Cambridge

Howson, A.G. & Wilson, B. (Eds.): 1986, *School Mathematcs in the 1990s*, ICMI Study Series Cambridge University Press, Cambridge

Mellin-Olsen, S.: 1990, 'Oppgavediskursen', in Nissen, G., & Bjørneboe, J. (eds.): *Matematikundervisning og Demokrati*, Statens Humanistiske Forskningsråd, Initiativet vedrørende Matematikundervisning, RUC, 47–64.

Nesher, P. & Kilpatrick, J. (Eds.): 1990, *Mathematics and Cognition: A Research Synthesis by the International Group for the Psychology of Mathematics Education*, ICMI Study Series, Cambridge University Press

Niss, M.: in press, 'Assessment of Mathematical Applications and Modelling in Mathematics Teaching' to appear in the proceedings (edited by Jan de Lange et al.) of the Fifth International Conference on the Teaching of Mathematical Modelling and Applications, ICTMA-5, Nooordwijkerhout, the Netherlands, 9–13 September 1991.

Niss, M. (Ed.): 1992, *Cases of Assessment in Mathematics Education: An ICMI Study*, Kluwer Academic Publishers, Dordrecht.

Mogens Niss
IMFUFA,
Roskilde University
Denmark

JEREMY KILPATRICK

THE CHAIN AND THE ARROW: FROM THE HISTORY OF MATHEMATICS ASSESSMENT

"Al niño hásele de proveer de materia, para que se revele el ingenio con su ácción y reacción, pues no hay cosa alguna cuya naturaleza se puede definir si está in inactividad total. La primera preuba a que sometia Pitágoras a sus iniciados era la Aritmética para explorar y averiguar su penetración. No hay cosa que mejor descubra la agudeza mental, como la expedicìòn en el arte de hacer cuentas y la cortedad de entendimiento la torpeza en el contar."

Juan Luis Vives (1531/1947, p. 567)

"¡Oh quién entrara hoy día en las escuelas de nuestros tiempos haciendo cala y cata de los ingenios!

¡A cuántos trocara las ciencias y cuántos echara al campo por estólidos e imposibilitados para saber!

¡Y cuántos restituyera de los que por tener corta fortuna están en viles artes arrinconados, cuyos ingenios crió Naturaleza solo para letras!"

Juan Huarte de San Juan (1575/1989, p. 224)

1. INTRODUCTION

The problem of assessing the mathematics pupils have learned has inevitably been intertwined with the questions of who should receive additional mathematics instruction and how that instruction should be managed. Scholars appear to have begun the empirical inquiry into these questions from the perspective of psychology during the Renaissance, asking what mental abilities are, how they develop, and how they are differentially disposed across people and across the requirements of various disciplines.

One of the first to address these questions was the humanist Juan Luis Vives, who distinguished four functions of cognition, i.e., of the "inner senses": "the function of receiving the sense images (imagination), the function of retaining them (memory), the function of combining them (fantasy), and finally, the function of assessing them (estimative)" (Noreña, 1970, p. 258). Cognition, together with sense perception, physical characteristics, and emotional nature, comprised one's disposition. Vives believed that education should be adapted to the various dispositions pupils bring to school, and he proposed a "trial of wits" as a means of judging a pupil's

M. Niss (ed.), Investigations into Assessment in Mathematics Education, 31–46.

intellectual powers and character. He took the revolutionary position that pupils should be given individual guidance and instruction (Noreña 1970, p. 192). A pupil who had been accepted into school should be observed for some months at play and in conversation, with the resulting personal record being reviewed at intervals by faculty members (Vives, 1531/1947, p. 568).

Vives' proposal of the trial of wits probably influenced his compatriot Juan Huarte de San Juan (1575/1989) to write his *Examen de ingenios para las ciencias* [Examination of Wits] (cited in Watson, 1913, p. cxix). In this work, which has been termed "an early essay in behaviorism" (Read, 1981, p. 127), Huarte, like Vives, was concerned with matching pupils' skills to the disciplines they would pursue. Huarte regarded the three faculties of mind as memory, understanding, and imagination, with memory best suited to arithmetic, understanding to the physical sciences, and imagination to higher mathematics (Read, 1981, pp. 89–90). He was more of a rationalist and natural philosopher than a humanist, seeing the pupil as someone possessing natural abilities that required stimulation and cultivation more than direct instruction from a teacher. Neither Vives nor Huarte attached much importance to mathematics as a practical subject, but both saw it as demanding reasoning as well as memory.

In their view of cognitive abilities as being ordered according to degree of excellence, with reasoning and creativity ranking well above memory and sense imagery, Vives and Huarte were reflecting the prevalent belief that all of nature, the faculties of the mind included, must be hierarchical (Read, 1981, p. 60). The universe is ordered according to the Great Chain of Being (Lovejoy, 1936/1964), with God at the top, Hell at the bottom, and humans in between, distinguished from and superior to the other animals by virtue of their higher cognitive powers. Similarly, by contending that because different disciplines demand cognitive abilities in different configurations, students ought therefore to be selected for and assigned to instruction according to the abilities they possess, Vives and especially Huarte were propounding a rationalistic view of schooling. The "hidden *arrow of rationality*" underlies the causal link they saw between assessment and subsequent instruction. Assessment was to become the tool for deciding who would study which subjects for how long, the tool for rational decision making in education.

Mathematics played a key role in both visions. With content that extended from counting to theorem construction and proof, it displayed the range of intellectual demands required for a view of cognition that was graded into levels. As a challenging and difficult subject in the curriculum of most schools, it provided an excellent reagent for testing pupils' fitness to be taught. In the modern history of assessment in mathematics, one can see the powerful influences of the twin ideas that mathematics is a hierarchically organized school subject and also a useful vehicle for regulating education.

2. HIERARCHICAL MENTAL ABILITIES

The science of mental testing dates to 1869, when Francis Galton, in his *Hereditary Genius*, began to apply Darwin's theory of evolution to the study of psychology. Galton, like James Cattell and other 19th century psychologists who followed, used tests of simple responses such as discrimination and reaction time, but little evidence emerged that such responses were related to mental prowess (Carroll, 1978, p. 6). The giant step toward relating mental development to the performance of standard tasks was taken after the turn of the century by Alfred Binet, who along with Théodore Simon, discovered how to assess mental ability by using "complex" tasks that were closer to what one might encounter in everyday life (Carroll, p. 7). The Binet-Simon intelligence test, which was to influence the construction of numerous subsequent tests, put mental ability on a scale from low to high. Tasks were arranged in increasing order of difficulty, and the examinee received a score based on the point at which successful performance began to be outweighed by unsuccessful performance. Subsequent group tests changed the mode of scaling to yield scores based on the number of correct answers, but the notion of a scale remained.

In 1905, the same year in which the Binet-Simon test was published, Binet opened the first French laboratory school for experimental pedagogy in Paris. He encouraged the principal of the school, his friend and collaborator V. Vaney, to produce achievement tests that might serve as "barometers" of instruction (Wolf, 1973, p. 170). Vaney's first test assessed achievement in arithmetic and contained items that "average" pupils at each Grade from 1 to 7 should be able to do successfully. Binet and Vaney used the test results to help identify retarded children. This pioneering work in achievement testing may have helped Binet break free from the dominant view that different mental faculties should be studied separately, leading him to rank tasks empirically according to their difficulty and apart from their content (Wolf, pp. 171–176). In the years since, debate has raged as to the nature of intelligence and what is being measured by intelligence tests (see Carroll, 1979; Gould, 1981), but the idea of scaling tasks by difficulty as an indicator of mental prowess has remained relatively unchallenged.

Achievement testing in the United States began as early as 1845, when:

"The Boston School Committee, proud of what their schools were accomplishing and under pressure from Horace Mann, Secretary of the State Board of Education in Massachusetts, to show that the schools were doing the job the state expected of them in view of the generous financial appropriations they had been given, undertook a comprehensive school survey." (Kilpatrick, in press).

The examiners published a table ranking the writing and grammar schools of Boston in order of the average achievement of a sample of their pupils on a series of written tests. Although they were quick to point out that the rankings gave only approximate measures of the pupils' intellectual accomplishment and were inadequate measures of a school's merits (Caldwell & Courtis, 1925, pp. 180–181), that did not stop school administrators from using achievement testing to compare schools and teachers. By the 1870s, written achievement tests were widely used in large cities and certain states, with the results occasionally published in newspapers (Tyack, 1974, pp. 47–49).

Educators began to worry about how these tests were being used and began to request standardization. Standardization usually meant that information would be available about the distribution of performance on the test by some reference group but might also mean that instructions for test administration would be specified, the test would yield easily scored responses, and advice would be given on how to interpret the results. If tests were going to be used for comparison purposes, some standardization seemed essential. An early attempt to scale achievement test scores and provide norms for each grade level — following the path of Binet and Vaney — was made by Edward L. Thorndike and his students, who developed achievement tests in arithmetic and other subjects from 1908 to 1916.

"By the time that America entered World War I, more than two hundred achievement tests were available for use in the primary and secondary schools ... [with] eleven available in arithmetic alone" (Resnick, 1982, p. 181).

Apart from issues of test construction and standardization, there was also the issue of how tests related to the school curriculum. A leader in examining the last issue was Ralph W. Tyler, who in 1931 reported on his efforts to construct achievement tests for various university courses. His procedure was to formulate the course objectives and define each objective in terms of student behavior. Then he would collect situations in which students would reveal the presence or absence of the objective, present the situations to students, and evaluate their reactions in light of the objective. He would determine the objectivity and reliability of the evaluation and improve them if necessary as well as developing more practicable measurement methods if they were necessary. He claimed to have found eight major *types of objectives*: (a) information, (b) reasoning, (c) location of relevant data, (d) skills characteristic of particular subjects, (e) standards of technical performance, (f) reports, (g) consistency in application of point of view, and (h) character (Tyler, 1931/1934, pp. 6–7). At the time, Tyler neither linked these objectives to specific behavior nor arranged the behavior in order of complexity. By 1949, however, he had specified *seven types of behavior* — (a) understanding of important facts and principles, (b) familiarity with dependable sources of information, (c) ability to interpret data, (d) ability to apply principles, (e) ability to study and report results of study, (f) broad and mature interests, and (g) social attitudes. Furthermore, he claimed that each course objective had two aspects — content and behavioral — and he arranged the seven types of behavior along one dimension of a two-dimensional chart.

The next step was taken by Benjamin Bloom (1956) and his colleagues, who organized the objectives into a taxonomy (dedicated to Tyler) that attempted to reflect the distinctions teachers make, to be logical and internally consistent, to reflect current psychology, and to be both neutral and comprehensive. The taxonomy was designed to fit all school subjects. In the taxonomy, objectives were separated by "domain" (cognitive, affective, and psychomotor, with the cognitive receiving most attention), related to "educational behaviors", and arranged in ascending order from simple to complex: (a) knowledge, (b) comprehension, (c) application, (d) analysis, (e) synthesis, and (f) evaluation. Research evidence was cited to support the claims for the cognitive domain that "as the behaviors become more complex, the individual is more aware of their existence" (Bloom, p. 19) and that the individual abilities and skills to be found in the upper levels of the taxonomy are more efficiently learned than the knowledge at the lowest level (p. 42).

Since its publication, variants of Bloom's taxonomy for the cognitive domain have helped provide frameworks for the construction and analysis of many mathematics achievement tests (Begle & Wilson, 1970; Romberg, Zarinnia & Collis, 1990). Major studies such as the *National Longitudinal Study of Mathematical Abilities* and the *National Assessment of Educational Progress* in the United States, as well as the First and Second International Mathematics Studies, have arrayed test items and scales in content-by-process matrices in which the content dimension is some classification of mathematical topics and the process dimension is a version of Bloom's taxonomy.

The organization of test items into such matrices has tempted researchers to examine how items relate to other items in the same or different row or column. Items testing the same mathematical topic appear to be related, but items testing the same level of mental process do not always hang together. In the First International Mathematics Study (Husén, 1967, ch. 1), Robert Thorndike found striking differences within each country in the pupil's performance on items testing different content topics but only small differences between items classified at different cognitive levels. Correlations across countries between items measuring "higher" and "lower" mental processes suggested that "there appears to be some slight difference in the function being measured by the two scores, but the difference *is* slight, and the two scores are hardly different enough to make their separate analysis promising" (p. 36).

"Research studies that have shown differences between process levels have found such differences only within topics, and complexity of process has then been equated with difficulty of test item" (Hill, 1984, p. 227; Seddon, 1978).

The Bloom taxonomy has often been seen as fitting mathematics especially poorly. As Ormell (1974) noted in a strong critique of the taxonomy, Bloom's categories of behavior

"are extremely amorphous in relation to mathematics. They cut across the natural grain of the subject, and to try to implement them — at least at the level of the upper school — is a continuous exercise in arbitrary choice" (p. 7).

Additional criticisms have questioned the validity of the distinction between cognitive and affective objectives, the independence of content from process, and the meaning of objectives isolated from any context (Freudenthal, 1975; Kilpatrick, 1979). Nonetheless, the view of mental abilities and consequently of mathematical thinking and achievement as organized in a linear, hierarchical way has been powerful in 20th century assessment practice. It has deep roots in our history and our psyches (Romberg, et. al., 1990, White, 1977).

3. RATIONAL SOCIAL CONTROLS

About 2200 B.C., the emperor of China established the practice of testing his officials every third year (Du Bois, 1965). After three examinations, they were either promoted or dismissed. The civil service applicant in 1115 B.C. had to pass proficiency examinations in several fields, including arithmetic. By 1370 A.D., open competitive examinations were well established in China, and applicants who were successful in their written examinations at the district level went on to take more in the provincial capital and then, if successful, in Beijing. The system was not abolished until 1905, but by then it had already been praised and copied in Europe and the United States. Competitive examinations were adopted as a means of selection not only for the civil service but also for the military and the university.

The Rise of the Examination System

In the medieval university, the form of examination was the public *disputation*, in which students demonstrated mastery of their subject by using standard texts or authorities to argue for or against a set position. As a student at the University of Paris from 1509 to 1512, Vives saw how artificial and hollow the disputations had become, as students engaged not in a collective search for truth but a display of individual power (Noreña, 1970, pp. 190–191). In the schools, pupils wrangled, quarreled, and sometimes came to blows in imitation of university disputations. Vives argued forcefully against disputations, especially by young school children. He preferred "the silent and authentic confrontation of opinions within one's self" (Noreña, p. 190), as well as what today would be termed *peer tutoring*.

Nonetheless, at English and Irish universities in the 17th and 18th centuries, examinations remained primarily a mixture of disputation and catechism, with questions chosen from a set list and students expected to defend a position orally, using arguments that conformed to standard models (Foden, 1989; Hoskin, 1982; Morris, 1961).

The written examination in the European university appears to have developed at the beginning of the 18th century. Two major influences appear to have been the growing number of prize and scholarship competitions modeled on the Chinese civil service examination and the increasingly discredited but still widely used disputation. The admission of larger numbers of students to universities began to make oral examinations of their knowledge unwieldy. Students had to wait hours to be examined, and then the examinations were often rather perfunctory. Philip Ballard (1923) cites the testimony of John Scott, Earl of Eldon, who went to Oxford in 1766:

"An examination for a degree at Oxford was a farce in my time. I was examined in Hebrew and in history. "What is the Hebrew for the place of a skull?" I replied, "Golgotha". "Who founded University College?" I stated (though, by the way, the point is sometimes doubted) that King Alfred founded it. "Very well", said the examiner, "you are competent for your degree"." (p. 16)

More importantly perhaps, some fields of study appeared not well suited to the oral format. According to W. Rouse Ball (cited in Morris, 1961, p. 30), the first recorded written examination in Europe was a mathematics examination given by Richard Bentley, the new Master of Trinity College, Cambridge, in 1702. Clearly, it must have been difficult to examine students on their knowledge of Euclid's *Elements* without providing them any material for writing or drawing.

"Between 1770 and 1840 the growing importance of solving mathematical problems increased the emphasis [at Cambridge] on the written part of the examination, and the difficulty of weighing the oral as against the written parts led to the gradual abandonment of the disputation in favour of an examination based entirely on the more easily assessed written papers." (Morris, p. 31)

Oxford and other universities were following Cambridge's lead, particularly in mathematics and the sciences, but the oral examination retained its preeminence in many universities for many years and of course still survives in most universities in one form or another. Frank Foden (1989, pp. 6–20) describes the frenzied, complicated, and fatiguing oral examination procedure employed by Trinity College, Dublin, in the first half of the 19th century. Although James Booth, the originator of the system begun in the 19th century by which nonstatutory examining bodies set the curriculum and standards for much of English education, was a mathematician, he was also a graduate of Trinity College, Dublin, and always preferred oral to written examinations.

Booth's position lost favor, however, as England underwent an examination boom (Ballard, 1923, p. 2) in the mid-19th century. The government began in 1862 a scheme of *payment by results* (Howson, 1982, pp. 121–122; Morris, 1961, p. 8) in which money was granted to each school in accordance with the pupils' performance in reading, writing, and arithmetic. Individual oral examinations were not economically feasible without hiring enormous numbers of school inspectors, and so written examinations began to be used instead. Because all pupils in a class could sit for the same examination at the same time, written examinations were seen as offering efficiency, equal treatment, and freedom from examiner bias. At all levels of education, examinations gave society a mechanism for identifying talent and channeling it productively. Victorian England believed that "competitive, open, and increasingly written examinations would eventually remove the undesirable consequences of unregulated favouritism, and would, if universally extended, have a salutary influence on society generally" (Mac Leod, 1982b, p. 1).

A reaction against examinations set in near the end of the century. In 1888, some 400 eminent men and women signed an article in the *Nineteenth Century* protesting the "sacrifice of education to examination". "The protest concentrated on schools, but the universities were also held accountable for providing English education with 'a body but without a soul'" (MacLeod, 1982a, p. 12). Objections were raised to the compartmentalization of knowledge promoted by examinations and, in an echo of Vives's arguments against disputations, to their elevation of competition over understanding. The anxiety caused by examinations was distressing to some people, who saw the examination system as failing to promote the kind of permanent learning that educators ought to value.

In the United States at about the same time, school superintendents, college professors, and college presidents began to press for reform in the examination system. Americans looked at the English system and worried that they were headed in the same direction:

> "Our leading educators, as in England, have already learned the lesson that the examination system has been overdone. But lower down, in our vast system of common school education, the examination octopus is sending out its arms and grasping all that comes within its reach." (Ferree, 1889, pp. 1–2)

England seemed a poor example to follow. Robert W. Rogers (1889), a professor at Haverford College, observed that "it is a almost a truism, since the days of Matthew Arnold at least, that in Germany, the land of few examinations, scholarship has touched a mark far beyond that reached in England, where an examination begins everything and concludes everything" (p. 28).

For William Torrey Harris (1889), who was about to be appointed the United States Commissioner of Education, the evils in the English system arose because the examinations were controlled by outside boards rather than the professional teachers. An opposing view was advanced in 1890 by J. Rendell Harris, a professor at Haverford who had come from Cambridge University. He agreed that reform was needed, arguing that "the first thing to be reformed is the examiner" (quoted in Kandel, 1936, p. 33). He contended that the purpose of any examination is to reveal individual differences: "A well conducted examination divides the students one from another like the opening of a fan" (p. 33). A proper system would remove the practice of examining from teachers (who tended to be overly concerned about "democratic results") and put it into the hands of an external body of examiners, so as to "gather only the good fish into their appropriate vessels and throw the bad away" (p. 34).

Both the criticisms and the examinations themselves continued into the 20th century. Examinations were stressful and seemed not to reveal the examinee's understanding, true, but they were also easy and efficient means of deciding who had learned what the schools and universities were providing. As Arthur Schuster observed in 1905, "It is much easier to teach if you make the accumulation of knowledge the primary object, and it is so difficult to test by examination anything except the possession of knowledge" (quoted in MacLeod, 1982a, p. 14).

As examinations flourished in and outside schools and colleges, educators began to worry about the reliability and objectivity with which decisions were made about examination performance. (As early as 1888, Francis Y. Edgeworth in England had studied the sources of variation in competitive examination papers and how they might be controlled; see Kilpatrick, in press.) The science of educational measurement was seen as a means of providing a *new type* examination, one that would be more reliable and more objective to score than the typical essay examination of the time. In the United States, following the lead of Joseph M. Rice and E. L. Thorndike, educators began to use short answer, true-false, and multiple-choice tests (see Kilpatrick, in press, for a discussion of the measurement movement in arithmetic). Meanwhile, in Europe and especially in England, concern arose about what examinations were testing and whether they were being graded fairly. In 1910, Philip Hartog proposed in a speech at the Royal Society of Arts that an inquiry into examinations be undertaken. A committee was appointed, but funds did not permit the proposed inquiry to be pursued at that time (Monroe, 1939, p. 95).

In 1925, a joint commission of the Association of College Presidents of Pennsylvania and the State Department of Public Instruction of Pennsylvania began to consider the relationship between secondary school and college curriculums. They wanted to find out how pupils might be better prepared for college and how colleges might build on their students' previous achievements. They invited the Carnegie Foundation for the Advancement of Teaching to undertake a study, which became known as the Pennsylvania Study, of students' abilities and achievements as they proceeded from secondary school through college (Kandel, 1936, pp. 133–140; Travers, 1983, pp. 135–142). The study proceeded from the premise that, whereas in Europe students' intellectual growth was expected to be systematic and continuous along distinctive lines, in the United States the student's program was a patchwork of unrelated courses, with achievement being measured in credits accumulated rather than accomplishments examined. The proposal was to follow students through their educational careers with extensive examinations at entrance to secondary school, at the end of secondary school, and near the end of college. The examinations were to be new type tests whose focus was to be on knowledge rather than on "higher mental processes" because knowledge was held to be essential to thinking:

"A high score in this kind of test does not infallibly demonstrate the attainment of what we call a liberal education; but a low score does infallibly demonstrate a lack of liberal education, because it reveals the absence of the foundation upon which a liberal education must stand... One may have a flourishing tree without fruit, but one cannot have fruit without a tree; knowledge — ample and accurate knowledge — is the tree on which the fruit we call culture must grow." (M. McConn, quoted in Kandel, p. 140)

The Pennsylvania Study lasted for more than a decade and became a major influence on the development of standardized objective tests for use in school and college assessment.

The International Examinations Inquiry

Frederick P. Keppel, president of the Carnegie Corporation and a good friend of Thorndike, who had been a fellow doctoral student at Columbia University, was instrumental in supporting a number of studies of achievement measurement at the time, including the Pennsylvania Study (Travers, 1983, p. 136). In a conversation with Paul Monroe, who was director of the International Institute of Teachers College, Columbia University, Keppel indicated that he and the Carnegie Corporation were interested in "doing something for education in Europe, something which would affect several countries" (Monroe, 1939, p. 8). The result of that conversation was the International Examinations Inquiry, sponsored by the Carnegie Corporation, the Carnegie Foundation, and the International Institute, and conducted from 1931 to 1938 (Kandel, 1936, pp. 163–175; Monroe, 1931, 1936, 1939).

The first Conference on Examinations was held at Eastbourne, England, in May 1931 and was attended by 28 delegates from England, France, Germany, Scotland, Switzerland, and the United States (Monroe, 1931). The delegates included Cyril Burt, Percy Nunn, Charles E. Spearman, Godfrey Thomson, Jean Piaget, and E. L. Thorndike. A difference of opinion emerged, with the French and German delegates somewhat skeptical about the ability of objective tests to assess the *culture générale* or *allgemeine Bildung* that education should promote and the American and British delegates seemingly more dissatisfied with current practice and more eager to experiment with objective tests and new schemes for grading examinations. As Philip Ballard, an advocate for objective testing, noted:

"We in England have just begun to suspect the infallibility of the examiner. That has gone very much further in America. In America the old type of examination is not only under suspicion; it is under arrest. In England it is under suspicion, but in some countries, apparently, it is not suspected at all." (Monroe, p. 91)

National committees were set up in several countries, a variety of studies were conducted, and two additional conferences were held in Folkestone, England, in June 1935, and in Dinard, France, in September 1938, at which the results of the studies were reported and discussed (Monroe, 1936, 1939). The last conference drew 36 participants, including delegates from Finland, Norway, and Sweden; observers from China and the Netherlands were also present.

As it turned out, the German Republic was "more ready for experimentation in education than either France or England" (Kandel, 1936, p. 169) in part because of the overcrowding of German schools, universities, and professions and the perceived need to raise standards. Despite the findings of studies that showed considerable subjectivity in the marks given by examiners, the director of the French Inquiry, Auguste Desclos, remained unconvinced that objective tests could capture all of the qualities of mind sought by educators and worried that if essays and dissertations were abandoned in favor of tests, classroom practice would suffer (Monroe, 1936, p. 147). Whatever changes were made in examinations, the "efficiency of teaching" should not be sacrificed to the efficiency of measurement (Monroe, 1939, p. 317). Desclos attributed the failure by over 2,800 teachers and administrators out of 3,000 to respond to a questionnaire on examinations to their satisfaction with the current system (Kandel, p. 169). Philip Hartog, director of the English Inquiry, encountered a similar complacency: A university official asked for help in supplying examination scripts announced: "We think that here ... we know everything that there is to be known about examinations" (Monroe, 1936, p. 5).

The Second World War brought an end to the International Examinations Inquiry and to the plans made at the final meeting to set up a permanent organization to study examination systems around the world. It is difficult to judge how far various countries modified their examinations as a consequence of the Inquiry, but there seems little doubt that it helped establish objectivity and reliability as desirable goals in an examination system by bringing forcefully to people's attention the degree to which uncontrolled marking schemes yielded highly divergent ratings of the same examinees.

The Inquiry produced technical studies, but the deeper problems of examinations were not simply technical. In a summary paper given at the end of the last Inquiry conference, I. L. Kandel, of the International Institute at Teachers College, Columbia University, observed that the conflict in education facing each country lay between adjusting pupils to a given concept of liberal education and adjusting education to the given pupils' abilities (Monroe, 1939, p. 313). That conflict, as it related to examinations, could not be resolved by studies of marking schemes. Rather, it concerned each society's view of one of the themes enunciated by Paul Monroe in his opening remarks at the first conference: "Examinations *as a means of social control*" (Monroe, 1931, p. 3).

4. ASSESSMENT IN A NEW CONTEXT

"L'examen combine les techniques de la hiérarchie qui surveille et celles de la sanction qui normalise. Il est un regard normalisateur, une surveillance qui permet de qualifier, de classer et de punir. Il établit sur les individus une visibilité a travers laquelle on les différencie et on les sanctionne." Michel Foucault (1975, pp. 186–187)

As the idea of the Great Chain of Being gave way during the 19th century under the onslaught of natural science and the theory of evolution, so the idea that mental abilities are linearly and hierarchically ordered seems to be giving way in the last half of the 20th century under pressure from educational, psychological, sociological, anthropological, and philosophical research. To the extent that pupils can still be thought of as possessing distinct mental abilities, it is clear that such abilities arise from multiple sources and interact in complex and diverse ways:

"Our data from cognitive development, from psychoneurology, from human learning, and from memory studies tell us that humans have multiple knowledge systems, multiple systems of representation, multiple gnostic centers, multiple short-term and long-term memories, and multiple laws of learning." (White, 1977, p. 40)

The hierarchical view of abilities, and consequently of behavior and objectives in school, is seen as leading to unfortunate educational practices:

"The idea that knowledge must be acquired first and that its application to reasoning and problem solving can be delayed is a persistent one in educational thinking. "Hierarchies" of educational objectives, although intended to promote attention to higher order skills, paradoxically feed this belief by suggesting that knowledge acquisition is a first stage in a sequence of educational goals. The relative ease of assessing people's knowledge, as opposed to their thought processes, further feeds this tendency in educational practice." (Resnick, 1987, pp. 48–49)

The construct of ability itself is coming under attack from researchers and theorists who are concerned about how an ability is to be assessed. When children given an assessment task to do are observed making a mathematical error, for example, the error can be understood as having occurred not because the children lack an "ability" but because they have not entered the social practice of the situation. When they are successful, it is because they have correctly read the cues and clues provided by the social interaction (Walkerdine, 1988, pp. 52–53). Under this interpretation, cognition cannot be studied apart from context; it is situated socially and culturally. We are urged to revise our conception of *task* to include "its source, social constraints, and multiple perspectives" (Newman, Griffin & Cole, 1989, p. 16).

The idea that a social practice such as school examinations can be regulated in a rational fashion has also come into question in the last half of the 20th century. The *purposeful-rational* model of social change (Popkewitz, 1988) suggests that people simply need to spell out their objectives and then engineer the social practice by following a sequence of steps until the objectives are met. That model is being abandoned, however, as policy makers discover that one cannot assume the context of a social practice can be controlled and the consequences of a social action foreseen. "Our social conditions contain a host of elements that interact in ways that are never fully specified, predetermined, anticipated, or willed" (Popkewitz, pp. 245–246).

At the final Conference on Examinations of the International Examinations Inquiry, Desclos posed a prophetic question:

"Is there not a danger, if we bind ourselves to the ideal of first setting up measurements and then conforming our teaching to those measurements, that we shall bring down the level of what we are teaching or shall limit ourselves in our effort to teach on that higher level that we so eagerly strive to attain." (Monroe, 1939, p. 129)

He was assured that no one was proposing the abandonment of subjective types of measurement but only the addition of objective measures. Several times during the conference, the argument was repeated that objective tests, far from diminishing the curriculum or distorting teaching, would enable teachers to diagnose pupils' difficulties and individualize their instruction. Essay examinations, clinical observations, and the like would provide additional information to be entered into the pupil's record.

A half century later in the United States, objective tests are being accused of having driven other forms of assessment out of the schools, trivalized learning, and warped instruction (Resnick, 1987; Romberg, et al.,

1990). In other countries, questions are being raised about "unintended, undesired, and sometimes even dangerous side-effects ... in the current methods of assessment" (Commission Internationale de l'Enseignement Mathematique (ICMI) 1990, p. 198). The vision of examination systems as an efficient and just means of diagnosing abilities and allocating pupils to appropriate social roles has become badly tarnished.

The most confident and optimistic voice to be heard at the conferences of the International Examinations Inquiry was that of E. L. Thorndike, the ardent positivist and trailblazing psychologist. Today, he almost seems justified in his faith; everything that exists *can* be measured — in some fashion. The 20th century has produced an assessment practice in education that is dominated the world over by *psychometrics*, the measurement of psyche. The challenge for the 21st century, as far as mathematics educators are concerned, is to produce an assessment practice that does more than measure a person's mind and then assign that mind a treatment. We need to understand how people, not apart from but embedded in their cultures, come to use mathematics in different social settings and how we can create a mathematics instruction that helps them use it better, more rewardingly, and more responsibly. To do that will require us to transcend the crippling visions of mind as a hierarchy, school as a machine, and assessment as engineering.

NOTE

I am grateful to Kenneth Ruthven for the metaphor of the arrow to describe the rationalization of social action.

REFERENCES

Begle, E.G. & Wilson, J.W.: 1970, 'Evaluation of mathematics programs', in Begle, E.G. (Ed.), *Mathematics education* (69th Yearbook of the National Society for the Study of Education, Part 1, 367–404), University of Chicago Press, Chicago.

Bloom, B.S. (Eds.): 1956, *Taxonomy of educational objectives: The classification of educational goals. Handbook 1: Cognitive domain*, McKay, New York.

Carroll, J.B.: 1978, 'On the theory-practice interface in the measurement of intellectual abilities', in Suppes, P. (Ed.), *Impact of research on education: Some case studies* (1–105), National Academy of Education, Washington, DC.

Caldwell, O.W. & Courtis, S.A.: 1925, *Then and now in education, 1845–1923: A message of encouragement from the past to the present*, World Book, Yonkers-on-Hudson, NY.

Commission Internationale de l'Enseignement Mathématique (ICMI): 1990, 'Assessment in mathematics and its effects: Discussion Document', *l'Enseignement Mathematique*, **36**, 197–206.

DuBois, P.H.: 1965, 'A test-dominated society: China, 1115 B.C.–1905 A.D.', in Harris, C.W. (Ed.), *Proceedings of the 1964 Invitational Conference on Testing Problems* (3–11), Educational Testing Service, Princeton, NJ.

Ferree, B.: 1889, March, 'Introduction. In Examination and education [American supplement]', *Nineteenth Century*, **25**, 1–2.

Foden, F.: 1989, *The examiner: James Booth and the origins of common examinations (Leeds Studies in Adult and Continuing Education)*, University of Leeds, Leeds.

Foucault, M.: 1975, *Surveiller et punir: Naissance de la prison [Discipline and punish: The birth of the prison]*, Gallimard, Paris.

Freudenthal, H.: 1975, 'Pupils' achievements internationally compared — the IEA', *Educational Studies in Mathematics*, **6**, 127–186.

Galton, F.: 1869, *Hereditary genius: An inquiry into its laws and consequences*, Appleton, New York.

Gould, S.J.: 1981, *The mismeasure of man*, Norton, New York.

Harris, W.T.: 1889, March, [Untitled], in *Examination and education [American supplement]*, *Nineteenth Century*, **25**, 22–24.

Hill, P.W.: 1984, 'Testing hierarchy in educational taxonomies: A theoretical and empirical investigation', *Evaluation in Education*, **8**, 179–278.

Hoskin, K.: 1982, 'Examinations and the schooling of science', in MacLeod, R. (Ed.), *Days of judgement: Science, examinations and the organization of knowledge in late Victorian England* (213–236), Nafferton Books, Driffield, UK.

Howson, A.G.: 1982, *A history of mathematics education in England*, Cambridge University Press, Cambridge.

Huarte de San Juan, J.: 1989, *Examen de ingenios para las ciencias [Examination of wits]* (Serés, G., Ed.), Ediciones Cátedra, Madrid (Original work published 1575).

Husén, T.: 1967, *International Study of Achievement in Mathematics* (Vol. 2), Wiley, New York.

Kandel, I.L.: 1936, *Examinations and their substitutes in the United States* (Bulletin No. **28**), Carnegie Foundation for the Advancement of Teaching, New York.

Kilpatrick, J.: 1979, 'Methods and results of evaluation with respect to mathematics education', in Christiansen, B. & Steiner, H.G. (Eds.), *New trends in mathematics teaching* (Vol. 4, 162–179), UNESCO, Paris.

Kilpatrick, J.: (in press), 'A history of research in mathematics education', in Grouws, D.A. (Ed.), *Handbook of research on mathematics teaching and learning*, Macmillan, New York.

Lovejoy, A.O.: 1964, *The great chain of being: A study of the history of an idea*, Harvard University Press, New York (Original work published 1936).

Monroe, P. (Ed.): 1931, *Conference on examinations* (Eastbourne, England, May 23–25, 1931), Columbia University, Teachers College, Bureau of Publications, New York.

Monroe, P. (Ed.): 1936, *Conference on examinations* (Folkestone, England, June 8–10, 1935), Columbia University, Teachers College, Bureau of Publications, New York.

Monroe, P. (Ed.): 1939, *Conference on examinations* (Dinard, France, September 16–19, 1938), Columbia University, Teachers College, Bureau of Publications, New York.

Morris, N.: 1961, 'An historian's view of examinations', in Wiseman, S. (Ed.), *Examinations and English education* (1–43), Manchster University Press, Manchester.

MacLeod, R.: 1982a, 'Introduction', in MacLeod, R. (Ed.), *Days of judgement: Science, examinations and the organization of knowledge in late Victorian England* (3–23), Nafferton Books, Driffield, UK.

MacLeod, R.: 1982b, 'Preface', in MacLeod, R. (Ed.), *Days of judgement: Science, examinations and the organization of knowledge in late Victorian England* (1–2), Nafferton Books, Driffield, UK.

Newman, D., Griffin, P. & Cole, M.: 1989, *The construction zone: Working for cognitive change in school*, Cambridge University Press, Cambridge.

Noreña, C.G.: 1970, *Juan Luis Vives*, Martinus Nijhoff, The Hague.

Ormell, C.P.: 1974, 'Bloom's taxonomy and the objectives of education', *Educational Research*, **17**, 3–18.

Popkewitz, T.: 1988, 'Institutional issues in the study of school mathematics: Curriculum research', *Educational Studies in Mathematics*, **19**, 221–249.

Read, M.K.: 1981, *Juan Huarte de San Juan*, Twayne, Boston.

Resnick, D.: 1982, 'History of educational testing', in Wigdor, A.K. & Garner, W.R. (Eds.), *Ability testing: Uses, consequences, and controversies. Part 2: Documentation section* (173–194), National Academy Press, Washington, DC.

Resnick, L.B.: 1987, *Education and learning to think*, National Academy Press, Washington, DC.

Rogers, R.W.: 1889, March, [Untitled], in *Examination and education [American supplement], Nineteenth Century*, **25**, 28–29.

Romberg, T.A., Zarinnia, E.A. & Collis, K.F.: 1990, 'A new world view of assessment in mathematics', in Kulm, G. (Ed.), *Assessing higher order thinking in mathematics* (21–38), American Association for the Advancement of Science, Washington, DC.

Seddon, G.M.: 1978, 'The properties of Bloom's taxonomy of educational objectives for the cognitive domain', *Review of Educational Research*, **48**, 303–323.

The sacrifice of education to examination: 1888–1889, *Nineteenth Century*, **24**, 617–652; **25**, 236–255, 284–322.

Travers, R.M.W.: 1983, *How research has changed American schools: A history from 1840 to the present*, Mythos Press, Kalamazoo, MI.

Tyack, D.B.: 1974, *The one best system: A history of American urban education*, Harvard University Press, Cambridge.

Tyler, R. W.: 1934, 'A generalized technique for conducting achievement tests', in Tyler, R.W., *Constructing achievement tests* (4–14), Ohio State University, Columbus (Original work published 1931).

Tyler, R.W.: 1949, *Basic principles of curriculum and instruction*, University of Chicago Press, Chicago.

Vives, J.L.: 1947, 'De las disciplinas', in Riber, L. (Ed. & Trans.), *Juan Luis Vives. Obras completas* (Vol. 2, 337–687), M. Aguilar, Madrid (Original work published 1531).

Walkerdine, V.: 1989, *The mastery of reason: Cognitive development and the production of rationality*, Routledge, London.

Watson, F.: 1913, *Vives: On education*, Cambridge University Press, Cambridge.

White, S.H.: 1977, 'The social implications of IQ', in Houts, P.L. (Ed.), *The myth of measurability* (23–44), Hart, New York.

Wolf, T.H.: 1973, *Alfred Binet*, University of Chicago Press, Chicago.

Jeremy Kilpatrick
Department of Mathematics Education,
University of Georgia,
USA

GEOFFREY HOWSON

THE RELATIONSHIP BETWEEN ASSESSMENT, CURRICULUM AND SOCIETY

1. INTRODUCTION

"If college lectures do comparatively little good without frequent examinations, such a practice is doubly necessary in a Mechanics' Institution. They must be conducted, however, in such a manner as not to alarm the timid, and should be as much as possible in the style of conversation between the lecturer and his pupils, trying to elicit whether he has been rightly understood".

This quotation is from an article describing attempts to supply evening education to working men which appeared in The Penny Magazine of December 13, 1834, a magazine for the general public. It illustrates two important, and long recognized, facets of assessment. First, the essential part which formative assessment should pay in teaching (although nowadays we take a broader view of this type of assessment than is exhibited in the quotation); secondly, the attendant dangers of dispirited and embarrassed students. The example is, I fear, typical; for, as we shall see, in almost every aspect, assessment procedures must attempt to find some middle path between what is accepted as essential and good and that which is regarded as dangerous and dispensable.

2. ASSESSMENT AND THE STATE

My task in this chapter is to consider the relations between assessment, curriculum and society. I should like to begin by concentrating on *assessment and society* — in particular, the state.

The first thing to note is that society is not particularly concerned with the formative aspects of assessment which, for the teacher and the educator, are the more essential ones. What society demands is *summative assessment* — as applied to individuals and in some countries, through the aggregation of student scores, to institutions and even educational systems. One might summarize the position by saying that *formative assessment* is part of education's control system, and summative assessment is often seen as part of that of society.

M. Niss (ed.), Investigations into Assessment in Mathematics Education, 47–56.

To understand the interplay between examinations (for that is the most relevant feature of assessment) and society, it is interesting to see what society has sought from them throughout their history.

3. ASSESSMENT IN HISTORY

The first formal examination system, with written papers, appears to have been that devised in China about 200 B.C. or possibly earlier. It served to govern entry into official positions and acted as

"(a) a screening service (to select for a future role),
(b) a provider of incentive,
(c) a stabilizer of membership of the power hierarchy,
(d) a maintainer of the tradition of scholarship". (Griffiths & Howson, 1974, p. 28).

This examination system had an on-off existence for many centuries, but influenced Korean and Japanese customs. For example, Shuang Chi, a Chinese official, travelled to Korea in 958 A.D. and advised the Koreans to introduce mathematics into their examination system. (Needham, 1954, p. 139).

Certainly then, China, Korea and Japan used formal assessment procedures many, many years before countries in the West. Although, universities began to be widely established in Europe from the end of the thirteenth century, their examination systems were rudimentary and frequently mere matters of form: attendance at lectures, participation in disputations, inconsequential oral examinations and, sometimes, the giving of sermons entitled one to a degree. It was, for example, not until 1750 that Cambridge introduced written examinations in mathematics and for a further century students at King's College there were excused sitting the examination. Residence in Cambridge University sufficed to qualify them for their degrees.

The famous Tripos examination served, however, to give Cambridge mathematics a new focus: It established "standards" and, through the complicated way in which students were classified and ranked, stimulated competition and a certain type of excellence. The lessons were not lost on others and in the following century examinations were to become a major tool of social policy in several countries.

In France, competitive examinations were seen as a means of creating an educational ladder which rewarded merit and gave no weight to patronage. The entrance examinations made the *Grandes Ecoles*, and entry to the senior positions of state, open to all.

The *baccalauréat*, and the Prussian *Abitur* became essential qualifications for state service as well as schools providing the sole key to university entrance. School leaving qualifications play similar certification roles today.

The advantages of written examinations, which allowed many students to be examined in an equitable manner and to be awarded marks in a way which was both reliable and also capable of being moderated, were argued by Lacroix (1805).

Here we note the importance which society assigns to the reliability of assessment procedures. Already, then, important differences can be discerned between the emphases of educators and society:

	Educators' desire	Society demands
Mode	formative	summative
Qualities	validity beneficence	reliability

Note: Beneficence refers to the "backwash" on teaching and learning of possible good and bad effects. How do we ensure that good effects predominate — that is, that the assessment is beneficent? It must also be observed that formative assessment operates at a variety of levels: it can affect individuals, classes, schools, ... or it can be designed to provide information for mathematics educators and so affect actual teaching at one remove.

Figure 1

In Britain, Adam Smith, the economist, argued that examinations fostered excellency and exertion because of the way they stimulated rivalry and ambition; John Stuart Mill advocated the compulsory examination of children every year (whilst simultaneously battling against state control of education and a national curriculum). Examinations could classify, certify, stimulate, and, as France had shown, replace patronage. Arguments were put, then, for screening all candidates for public employment. It was even proposed that Government ministers should not be allowed to take office until they had passed suitable written test (Griffiths & Howson, op cit, p. 34). The catastrophes in the Crimean War, brought about by incompetent generals who had bought their ranks rather than attained them through experience, merit, and knowledge, finally led to the widespread use of examinations for screening purposes in England.

Control of examinations means control of the curriculum. This is a very powerful social factor. In some countries, their introduction under state control was used to help wrest control from the hands of the churches. Nowadays, newly independent countries wish to demonstrate that independence by forming national examination boards to provide services previously supplied and directed by the former colonial powers. In England the universities grasped the initiative in the 1850s and asserted their influence over the curriculum in the grammar schools. Attempts were made to provide alternative examinations, having a more technical and vocational, and less academic bias, but these never achieved much success or acquired the needed prestige. In recent years the Government has tried to assume control of the examination system whilst still leaving its administration in the hands of the independent boards. In Prussia, control was devolved to teachers: a clear indication of their status in nineteenth-century Prussian society. Gert Schubring has suggested to me (see Howson, 1991) that it was the lower social status of the teachers in what are now the southern *Länder* which led to their Abitur being more centrally controlled — the teachers lacked the status to take responsibility for certification of pupils. This example hints at another important aspect of the interaction between assessment and society: who can be entrusted with powers of certification and the responsibilities these bring with it?

Not only were examinations taken up with enthusiasm in France and England, but they were widely exported throughout their Empires: "Examinations have been the bane of Indian education" (Archer, 1921). If they were a bane in India, what can be said about their influence in Lesotho as demonstrated by the following statistics?

	Number of candidates obtaining grade						No.	No.	%
Year	A	B	C	D	E	U	Sat	Pass	Pass
1973	1	27	51	50	70	342	541	199	36.8
1974	3	20	71	39	75	630	838	208	24.8
1975	1	23	91	67	125	975	1282	307	23.9
1976	1	17	47	56	66	819	1006	187	18.7

Table 1 *Cambridge Overseas School Certificate in Mathematics 1973–1976 Lesotho*

Lesotho is perhaps an extreme case and one hopes matters have improved in the last few years, but what can be said in favor of an examination system, designed for students elsewhere, which labels over 80 percent of a developing country's élite (for there is much selection and dropping-out prior to the taking of School Certificate) as failures? A system which provides rewards and stimulation must also bring with it failure and dispiritment. It may show a few students to be the equals of their English counterparts, but it merely demonstrates that many, for perfectly good reasons, are not.

4. EXAMINATIONS AND SOCIAL CHANGE

What of the examination as the key to ensuring that all have equal chances? Dhombres (1987, p. 13ff) tells how in the very early days of the

École Polytechnique, 60 percent of students came from humble origins whereas only 14 percent had monied backgrounds. Only fifteen or so years later, by 1815, over 52 percent of the students came from the "couches supérieures", the upper layers, of French society. Soon entry was almost entirely the preserve of students from well-to-do families. Subsequent social swings make interesting reading.

In a similar fashion, changes made in the nineteenth century to entrance scholarships and awards at Cambridge University, so that they were open to all on "merit", resulted in "the erection of new and impassable class barriers which prevented the entry of the poor scholar" (Simon, 1960, p. 299). The two premier English universities became the preserve of the middle and upper classes who could be provided with the schooling which allowed them to demonstrate their merit. To this day independent schools send more pupils to Oxford and Cambridge than do state schools — even though the latter are far more numerous. Attempts to make the system more "meritocratic" are largely self-defeating. As was observed of education in general, "one might even define middle-class in this context as ability to profit from the educational system!" (Griffiths & Howson, op cit, p. 31).

Another interesting way in which examinations can result in a "closed shop" discouraging social mobility is provided by Kim (in press). In the list of successful candidates for the Korean mathematics examination in the 400 years from 1400 to 1800 there are 1,627 names. Their fathers' occupations were given as herbalists, 124; translators, 75; astronomers 6; and mathematicians, 1,422.

The examination system, then is a blunt instrument for social change. It can allow the gifted child to prosper — I personally owe a lot to it — but for the bulk of children it does little.

Even entry to the gateway can be limited: for example, objections were raised early in the nineteenth century that the Abitur was only open to students from the *Gymnasiums*.

The problems of a meritocratic educational ladder have been clearly recognized on two major occasions this century. In September 1918, Russia removed all university entrance requirements. Examinations and grading were also abolished: They led, so it was claimed, to formalism in teaching and discouraged creativity amongst pupils (Castles & Wüstenberg, 1979, p. 48). Somewhat similar actions were taken as part of the Chinese Cultural Revolution. Mao Tsetung attacked the Chinese educational system in which "teaching is by injection instead of through imagination" and "the method of examination is to treat candidates as enemies and ambush them" (Ch'en, 1970, pp. 21–22). In both cases, however, examinations soon made a comeback.

The place of examinations and educational ladders within socialist societies through the years is an interesting one which merits more attention than I can give it. Here I have only been able to hint at some of the ways in which examinations affect individuals within a society. There are, however, other influences both on society and on the way mathematics is taught.

5. THE ASSESSMENT OF EDUCATIONAL SYSTEMS

First, we must mention the use of assessment as a measure, not of a student's attainment, but of the way in which a school or a system is attaining. In 1862, the English politician, Robert Lowe, introduced a scheme which was aimed to ensure that grants to elementary schools would be related "to the attainment of a certain degree of knowledge by the children in the school". Certain "standards" were laid down in reading, writing and arithmetic, e.g., Standard 2 was to carry out "a sum in simple addition or subtraction, and the multiplication table". (In the recent English National Curriculum such work is subdivided between "Levels" 2–4 — plus ça change ...!) Schools were granted 8 shillings per year (40p) for each scholar who attended more than two hundred mornings or afternoons, but 2s 8d (13p) was deducted for each subject in which (s)he failed to reach the standard. According to Lowe it was a system which "if ... not cheap, shall be efficient: if it is not efficient, it shall be cheap". As he argued elsewhere:

"it is [not] part of the duty of the Government to prescribe what people should learn, except in the case of the poor, [for these] we must fix upon a few elementary subjects to get anything done at all ... The lower classes should be educated to discharge the duties cast upon them ... The higher classes ought to be educated in a very different manner in order that they may exhibit to the lower classes that higher education to which, if it were shown to them, they would bow down and defer" (quoted in Curtis, 1967, p. 256)

Such attitudes have, of course, changed, yet we note that in England independent (private, fee-paying) schools are still not obliged to follow the National Curriculum.

The outcomes of Lowe's "payment by results" scheme were mixed. As so often happens with "reform by examination", gains were registered initially, but then those interested, whether middle-class parents, students, or teachers, learn to cope and to divert the system to their own ends. As the leading inspector of schools, Matthew Arnold reported in 1870 (Howson, 1982, p. 121), "in the game of mechanical contrivances the teacher will in the end beat us ... by [getting] children through the ... examination ... without their really knowing [anything] of these matters". Children "practiced [sums] all the year round" but were not taught arithmetical principles "or introduced into the science of arithmetic". Although assessment techniques are now more sophisticated, so are the attempts to corrupt and defeat them!

There are successes to be listed, particularly in the field of the introduction of new content (more than counterbalanced, one suspects, by the failures), but little of lasting value appears to have been achieved by this means alone. Improvements in learning come from better teachers. Good examination practices and the opportunities these present can encourage teachers to develop. By themselves, though, changes in examination procedures are likely to achieve little.

The use of aggregated student scores to measure the performance of an educational system or for comparative purposes is a recent phenomenon. Interest in the education systems of other nations grew rapidly in the early eighteenth century. Reports were published on visits to France, England, Prussia, the U.S., and, of course, the teacher-training establishments of Switzerland. *ICMI* (the *International Commission on Mathematical Instruction*), itself, came into being in response to a desire to share information about practices and experiences. However, the realization that ever larger shares of national budgets were being devoted to education, brought new market considerations to bear. The total expenditure world-wide on mathematics education is colossal. It must account for roughly 10 percent of all national educational budgets (taking into account overheads), and expenditure on education might well amount to 10 percent of a country's budget. It is only to be expected, then, that increasingly governments will want to know whether educational systems are providing taxpayers with a reasonable return for their investment.

One initiative which resulted from such demands was the *Assessment of Performance Unit*, established in 1977 in the UK, "to promote the development of methods of assessing and monitoring the achievement of children at school and to seek to identify the incidence of underachievement". The aim was to make generally available a national picture of some aspects of school performance. The unit, however, was not concerned with the performances of individual children, schools or local education authorities (school boards). This latter consideration made the scheme acceptable under a Labour Government.

International comparisons began in the 1960s with the IEA's *First International Mathematics Study* and have continued through SIMS and the work of the *IAEP* (*International Assessment of Educational Progress*), along with many smaller, but nonetheless, valuable studies. It is relatively easy to criticize aspects of the work of all these major international studies, to describe their testing procedures as limited and simplistic, and to deplore some of the uses to which the data they have supplied have been put. Nevertheless, it must be accepted that such international comparative studies are here to stay, and rightly so. They offer much of educational value and, of course, increasingly there will be pressures on countries to participate and for educational systems to demonstrate exactly what goods, and in what amounts, they are delivering. One might deplore the metaphor, but as I have indicated, enormous sums of money are involved and politicians will demand accountability of some sort.

6. THE PRESTIGE OF MATHEMATICS

Mathematics occupies a particularly prominent position in comparative educational considerations. First, because the existence of a large "canonical" core within school mathematics means that the subject lends itself to such comparisons (imagine what would happen to an International History Study). More importantly, though, it is viewed as a necessary ingredient of the education of both an intellectual élite and of a technically-informed workforce. Mathematics is seen as an essential prerequisite for technological and scientific advancement. Mathematicians have been telling governments this for many years and we cannot complain if, as a result, we are now called to account for our work.

Complaints about the desirability of such studies would not, then, appear to be particularly relevant. They exist and they are likely to continue to exist. Our task is to make them as effective and as beneficial as possible.

The prestige of mathematics, and of attaining qualifications in the subject, is accompanied by considerable dangers. Let me quote André Révuz, writing of the position in France in the 1970s:

> "An educational stream is regarded as of more or less value in accordance with its level in mathematics ... Success in mathematics has become the quasi-unique criterion for the career choices and the selection of pupils. Apologists for the spread of mathematical culture have no cause to rejoice at this; this unhealthy prestige has effects diametrically opposed to those which they wish for. The cause of this phenomenon is ... that society ... has a need for a selective system which runs counter to attempts at democratization ... Many mathematics teachers allowed themselves to be caught up in this ... process ... It is easier to turn away those who arrive in class with an unacceptably low attainment level ... than to try to set them back on their feet. Others, driven to despair by [this] divisive role ..., reach the stage of wanting to suppress the difficulties inherent in the subject and to present it at such a low level that it is accessible to everyone". (Révuz, 1979).

The article is a very fine one and Révuz goes on to show how this latter attitude is detrimental to both the welfare of the students and that of the subject. But what is to be done?

Here we see another way in which assessment interacts with society and the curriculum. What happens if the flow of students into the different branches of higher and further education is unbalanced, or if "society" demands, as in France and England, that there should be increased access to parts of education traditionally restricted by means of examinations? True we can have a *baccalauréat* or A-level for all, but at what cost and with what repercussions? The dilemma of "they all shall have prizes" was remarked upon by Lewis Carroll. G. H. Hardy argued that the only examinations that did no harm were those that were so simple that everyone passed. But do they do any good? Again, there is a delicate balance to be found.

7. ASSESSMENT AND CURRICULA

I have little space to comment specifically on the relations between assessment and curricula. In some ways this is an "internal" matter of which we are all aware. No sooner had examinations been introduced than they became the targets for criticism. They "sat like a blight on education" and forced teachers to neglect what was educationally in the best interests of their pupils in favor of work directed specifically to the examination. Tocqueville, the French historian, warned the English in 1855 against following French examination customs: they were incompatible with the English constitution. (In a similar manner, the English educator, J. S. Howson, argued in 1859 that girls because of their "more excitable and sensitive constitutions" should not be exposed to examinations). Tocqueville argued that stereotyped forms of examination reacted upon and in some measure governed courses of study, uniformity of examinations leading to uniformity of study. There was a tendency "almost inevitable in schemes of examinations to push up standards", and excessive importance was placed on the holding of diplomas, "so that the want of them would come to be a barrier to entrance into trade, etc". (Roach, 1971, p. 85). As we have seen the argument about standards is two-edged. In some instances examinations do help push up standards. However, as Révuz indicated, in other cases, especially with the democratization of education, the reverse occurs. What cannot be denied, though, is that examinations have come to dominate the curriculum. Indeed, the new national curriculum in England is completely assessment led. First, attainment targets were set out and then the programs of study were defined to be that which was to be assessed.

8. THE DANGERS OF ASSESSMENT

During my time as a conscript in the Royal Air Force, I was taught "if it moves salute it". Nowadays, in England, and I fear in many other countries, the message has become, "if it moves assess it". There is no doubt that assessment is an essential part of teaching. There is also no doubt that we as educators must meet society's demands for summative assessment of students and for formative assessment of teachers, schools and educational systems. What must be done, though, is to ensure that in all cases the beneficial outcomes outweigh the detrimental ones. I am not convinced that at the moment this is the case. Let us hope that one outcome of today's attention to assessment issues will be an identification of ways in which assessment procedures can be improved so as to enhance, rather than to constrain, the quality of teaching and learning. It is also essential that we should consider how the general public can best be helped to understand the strengths and weaknesses of assessment and examinations, and that they come to realize that a surfeit of something, which in small doses can be good, may be damaging or even fatal.

Finally, I should like to draw to your attention a comment on examinations which I discovered when preparing this chapter. It was made by the famous mathematical-physicist, J. J. Thomson, who himself benefitted greatly from the meritocratic educational ladder: examinations are "a presumptuous attempt to gauge the depths of human ignorance" (Ballard, 1923, p. 28). As Antoine Bodin pointed out at the ICMI Study Conference, it is essential to ensure that present-day attempts to gauge human knowledge are not equally "impudent, presuming [or] over-confident".

REFERENCES

Archer, R.L.: 1921, *Secondary Education in the Nineteenth Century*, Cambridge University Press, Cambridge

Ballard, P.B.: 1923, *The New Examiner*, University of London Press, London

Castles, S. & Wüstenberg, W., 1978, *The Education of the Future*, Pluto Press

Ch'en, J.: 1970, *Mao Papers*, Oxford University Press, Oxford

Curtis, S.J.: 1967, *History of Education in Great Britain*(7th ed.), University Tutorial Press

Dhombres, J.: 1987, 'Introduction', in Fourcy, A. : *Histoire de l'École Polytechnique* (reprint), Belin

Griffiths, H.B. & Howson, A.G.: 1974, *Mathematics: Society and Curricula*, Cambridge University Press , Cambridge

Howson, A.G.: 1982, *A History of Mathematics Education in England*, Cambridge University Press, Cambridge

Howson, A.G.: 1991, *National Curricula in Mathematics*, Stanley Thornes, for the Mathematical Association

Kim, Y.W.: 'Tradition and Mathematics Education in Korea', *L'Enseignement mathématique* (to appear)

Lacroix, S-F.: 1816, *Essais sur L'Enseignement* (2nd ed.), Paris

Needham, J.: 1954, *Science and Civilization in China*, Vol. 1, Cambridge University Press, Cambridge

Révuz, A.: 1979, 'Changes in the teaching of mathematics in France', *Math. Gazette*, **63**, 241-50 (Original in French in *Educ. Stud. Math.*, **9** (2), 1978)

Roech, J.: 1971, *Public Examinations in England 1850-1900*, Cambridge University Press

Simon, B: 1960, *Studies in the History of Education 1780-1870*, Lawrence and Wishart

Geoffrey Howson
Department of Mathematics
The University of Southampton
England

JIM RIDGWAY & DON PASSEY

AN INTERNATIONAL VIEW OF MATHEMATICS ASSESSMENT – THROUGH A CLASS, DARKLY

1. INTRODUCTION: INTERNATIONAL SNAPSHOTS

A Focusing Question: *What Does an International View Reveal About Assessment?*

A view of assessment in any educational system is likely to stimulate ideas about the variety of functions which assessment serves. This paper begins with 7 snapshots of educational systems together with some observations. These snapshots should be viewed as archetypes, not case histories. Detail is deliberately restricted; the principal aim is to stimulate reflection, not to overload the reader with information. Much of the evidence is taken from Cameron, Cowan, Holmes, Hurst & McLean (1983).

Snapshot 1: Afghanistan

Afghanistan is Islamic, but otherwise linguistically and culturally diverse. There are tensions between Islamic culture and Western values. Participation rates in education in 1979 were amongst the lowest in the world. Eighty-five percent of the population was illiterate. Mass participation and the education of girls were seen to go against Islamic values; approximately 5 times more boys than girls were educated at each Grade level. Since 1979, pupils in each class have been graded from 1 (poor) to 5 (very good). An average of 2 or above allows a student to be promoted to the next Grade. Students have a right of access to secondary education, if their grades permit. Between Grade 1 and Grade 9, participation rates decrease at an average of about 20 percent per year. Girls do not study logic, and boys do not study psychology ... Included in Cameron's et al.'s list of Afghanistan's educational problems are: "a corrupt examination system relying mainly on memory work ... low student attendance rates ... academic [curricula] non-practical and irrelevant to the needs of the economy or even to the personal requirements of a predominantly agricultural population". Assessment is used for *minimum competency testing* in this system. It seems likely that educational reform will require rather more than an improvement in the quality of mathematical assessment.

M. Niss (ed.), Investigations into Assessment in Mathematics Education, 57–72.

Snapshot 2: Zambia

In Zambia, education is valued highly; primary education is free and compulsory. At Grade 7, students take a Primary School Leaving Examination which not only certifies the completion of primary education, but also determines who will receive access to secondary education. The examination comprises multiple-choice papers in English, mathematics, social studies, science and two languages, and also a test of verbal reasoning and nonverbal reasoning. The latter two tests are used in the selection process, but not for certification. Approximately 20 percent of the Grade 7 school population enter secondary school at Grade 8.

Here, assessment is used for *certification*, and as a barrier to access to further study, which is seen as legitimate by students and society. Presumably, multiple-choice tests are used because of their cheapness.

Snapshot 3: Argentina

Argentina, in common with very many other education systems, provides selective schooling at secondary level (at around Grade 6 or 7). Common divisions are into technical or academic, or perhaps agricultural or clerical or craft-based institutions.

Assessment in such systems is used to *partition* students. The underlying beliefs are that tests can determine what educational strand is best for pupils, that individual differences remain relatively stable over time, and that the human resource needs of society should be reflected in the education system.

Snapshot 4: Japan

Japanese students obtained very high scores on both the *First International Mathematics Study* (FIMS) (Husén, 1967) and the *Second International Mathematics Study* (SIMS) (Robitaille & Garden, 1989). Among 13 year olds, their lead over other nations was about one standard deviation; at age 18 years, the specialist mathematicians were about two standard deviations ahead of Western students, and the top specialists were almost three standard deviations ahead. The average Japanese students obtained approximately the same scores as students from the top 1 percent in other countries. These differences start early. At age 6 years, Japanese students are about 0.6 standard deviations ahead of their American counterparts (Stigler, Lee, Lucker & Stevenson, 1982).

Japanese society is characterized by cultural and socioeconomic homogeneity, relative to Western societies (Shimahara, 1985), and formal education is seen to be a major factor in social mobility. The *juku system* of after school tuition attests to the high status of academic success; so too does the longer school day and the longer school year, and the high percentage of time devoted to mathematics.

At age 14 years, Japanese children take entrance examinations which allow entrance to a range of senior high schools. Differences between schools are known locally, and competition to gain access to the "best" schools is fierce. Access to universities is also by entrance examination. Innate ability is considered far less important than it is in the West as a determinant of educational success; examinations focus upon academic content. Universities have different reputations for excellence, which is reflected in the employment opportunities for graduates, (several companies recruit predominantly from a few universities), and in terms of the ease of entrance (Shimahara, 1985). Once in employment, job mobility is low, and so entrance to a popular company is seen as a guarantee of a lifelong job.

Motivation to succeed at the examinations is high; since examinations are content based, hard work and careful revision are sensible strategies for students to employ. Almost 40 percent of the school population enters higher education, and so these competitive pressures affect most students. These achievements have considerable personal costs to many of the individuals concerned.

Snapshot 5: The USA
American students did not star in SIMS. American society is characterized by cultural and socioeconomic heterogeneity. Many of the best universities are private rather than state institutions, and charge substantial fees, thereby attenuating access of students from poorer backgrounds.

Entrance to state universities is gained largely via good scores on the *Scholastic Aptitude Test* (*SAT*). The use of an aptitude test to select for university entrance is made possible by a cultural belief in "intelligence", and a belief that aptitude can be divorced from competence, and long-standing traditions of testing and selection for admission to the armed forces and business. SAT does not require students to possess detailed academic knowledge to perform well. SAT is necessary, in part, because of the absence of national systems of assessment, and in part because of cultural beliefs that able, less advantaged students, must not suffer discrimination which can arise from poor schooling. There is little pressure for students to master their mathematics, if the basis of college admission is their "aptitude", rather than their competence.

The *National Commission on Excellence in Education* (1983) published *A Nation at Risk* and argued for assessment-driven reform. Forty-four out of 50 states now have state-mandated elementary and secondary school tests for students, usually linked to state-defined goals. These are usually based on reading, language, and mathematics. Only six states now require assessment in science and social studies. Few other nations do this.

Malkins (cited in Green & Mulhearn, 1989) reports that American teachers with experience of state-imposed testing of curriculum topics express very negative views about the prospect of such tests being introduced into the UK. The students' feeling of failure was cited as the main disadvantage of such schemes.

Snapshot 6: England and Wales 1950 to 1988
The examination system in England and Wales derives from the needs of "new" universities to demonstrate that their undergraduate entrants were as knowledgeable as those who entered old established universities such as Oxford and Cambridge (Mathews, 1985). The *Advanced Level Examinations* were used to gain access to Advanced Level courses. These examinations were externally set and marked, and so gave an "objective" view of students' achievements; they were therefore attractive to employers, and soon came to dominate the whole educational agenda. For a long period, England and Wales employed a system of selective education where students were tested on the basis of some combination of intelligence, verbal reasoning, and mathematical ability at age 11 years, to determine the school best suited to their educational needs. The two commonest forms of secondary school were grammar schools, which followed a largely academic curriculum, and secondary modern schools which emphasized craft skills. Two sorts of examination were followed: The *General Certificate of Education* (GCE) was most commonly taken in grammar schools and aimed at the top 30 percent of the ability range; in secondary modern schools, the *Certificate of Secondary Education* (CSE) was taken and aimed at the next 30 percent of the ability range. In both, each school subject was examined separately. Both examinations were set and marked externally. The CSE had a component of coursework set and assessed within the school but moderated by other schools; GCE was assessed entirely by examination. In mathematics, this was a timed, written examination. To allow comparability of the examinations, the top grade at CSE was judged to be equivalent to a pass at GCE, and attempts were made to ensure an adequate overlap of syllabuses between the two courses of study. As a result, about 80 percent of students followed a course of study designed to serve the needs of the top 20 percent of the ability range, i.e., those likely to pass GCE. The backwash was impressive — and dire. A Grade 5 at CSE "the performance just below that to be expected from a pupil of average ability" was awarded to a student who obtained about 20 percent of the available marks.

Here assessment was driven by the needs of higher education. The world of work used inappropriate education measures as the basis for selection.

A new *National Curriculum* is now being introduced (Department of Education and Science, 1989). Just one of its elements will be noted here. The *Task Group on Assessment and Testing* was established to offer advice on appropriate methods of testing which are affordable and easy to use. This last phrase caused many hearts to sink, including mine:

"... an emphasis on tests which are "affordable" and externally moderated is likely to result in paper and pencil tests which are easy to mark, and which are likely, therefore, to fail to assess practical skills, problem solving, investigative work, or group work, all of which are essential elements of mathematical curricula". Ridgway (1988).

These worries are being confirmed by current events. Assessment methods chosen on the basis of costs are likely to distort mathematics curricula.

Snapshot 7: An Educational Experiment; Sri Lanka 1972–1982
This snapshot is based on a study by Lewin & Little, reported by Little (1990).

Pre 1972. The English GCE Ordinary and Advanced examination level were used. In 1970 a coalition of socialist and populist parties came to power. They argued that the country was too dependent upon foreign aid and expertise, and needed to develop indigenous resources. The examination system was seen to be too heavily focused upon examinations, and placed too little value on the development of skills relevant to local business and the economy.

"Of the great deficiencies created in our societies by colonial rule nothing is so pervading as a lack of moral courage and strength to think beyond the intellectualism imposed on us by the imperial powers ... The preoccupation of us all with what occurs in developed societies has stifled our intellectuals, thinkers and innovators in education". (Secretary of the Ministry of Education, quoted by Little)

1971. Riots indicated the height of tension and concern about the slowness of political reforms. According to Little, one commentator estimated that 85 percent of those detained after the insurrection had the GCE Ordinary level qualification. The education system was seen as a source of political unrest.

1972. Ordinary and Advanced Level examinations were replaced by examination systems aimed at the needs of the majority of students. It had no links to any certification outside Sri Lanka.

1977. The United National Party, whose base is traditionally business and conservative elements, came to power with a huge majority. Their goal was to increase international links, and to increase educational standards and to ensure that qualifications were comparable with those from developed nations. Ordinary and Advanced Level examinations were reintroduced.

External examinations in developing countries can continue as vestiges of colonial power. They may well continue to act as agents of cultural imperialism. The role of assessment to facilitate international mobility has a double edge. For the affluent, it can offer access to universities world wide, and for the less well off it can allow access to skilled jobs around the world. From the society's viewpoint, it can lead to a drain of the educated workforce.

2. PEERING AT THE PICTURES

Case studies based on single countries can help our thinking about the nature of large scale educational change; international comparisons generate a range of hypotheses about the roots of successful mathematics teaching. Assessment relates to a wide range of social issues, such as access to further education and employment. It often is intimately linked to educational and economic planning. While a good deal can be learned from international comparisons, great care must be taken when making generalizations, because of the overarching social and political contexts. In reviewing working systems of assessment, it becomes clear that any consideration of assessment within a country needs to take account of the broader educational and social settings, and the social beliefs which underlie them. For most practical purposes, social and political beliefs are at least as important as research-based educational wisdom. In any country it is necessary to review the system within which mathematical assessment is set, the way in which assessment can be done, and the particular attributes that are measured. This theme is revisited later.

Beliefs about the source of mathematical attainment are important. A society that believes heavily in "intelligence" and "aptitude" is likely to engender a far more fatalistic approach to mathematical development than one which considers hard work to be a prime cause of success. A strong belief in "aptitude" is probably a serious barrier to effective mathematics teaching.

Political climate can promote the needs of the business and industrial sectors. Business-oriented societies are likely to emphasize efficiency and accountability of teachers and schools, external assessment, and international comparisons. International aspects of assessment are important; widely recognized standards might act to accelerate the process of redistributing the assets of poorer countries to richer ones. The "instrumental" assertion that students should work for qualifications in order to gain access to jobs is too simple. If jobs remain in short supply, and educational standards rise, then the entry qualifications will also rise.

While the relationship of the educational system to the social and political environment offers one direction for exploration, the relations of subsystems within the educational system offers another. Education systems differ in the extent to which they are "top down" or "bottom up" systems. The tightness of the coupling between top and bottom may determine the system, structure, and methods used for mathematical assessment. For example, school curricula can be distorted by the need of a few students to pass university examinations. More generally, if access to education is very limited, almost any kind of assessment can be used as a filter which is socially acceptable; if access is open, the role of assessment in specifying necessary component skills is more likely to emerge.

Access to technology offers seductively simple answers to very complex problems. Mark scanners make multiple-choice testing on a large scale very cheap and therefore attractive in comparison with other assessment systems. Their use is likely to corrupt broader educational goals. There is need for caution about the belief that the quality of mathematical education can be improved via better assessment methods. Far more significant educational reforms may be required.

Uses for Assessment

A wide range of uses for assessment emerge from peering at the educational snapshots. Uses include:

- explanation and maintenance of social divisions;
- qualification and access to higher education, or a career;
- prediction;
- selection, and plausible methods for rejection;
- certification of achievement, competence, or technical mastery;
- enhancement or reduction of personal status;
- enhancement or reduction of self esteem;
- driving or destroying attainment.

To this list can be added:

- providing feedback to students, parents;
- recruitment (e.g., to high performing schools);
- monitoring class progress;
- record keeping;
- identifying problematic aspects of mathematics;
- diagnosis;
- assessment of curricula, teachers, schools;
- setting behavioral objectives;
- definition of the curriculum;
- research into psychometrics.

These uses affect people at different levels within the education system in different ways. It is time to address some of this complexity in a more ordered way.

3. LAYERS OF ANALYSIS

The educational community can be viewed as a series of subcommunities, such as students, teachers, schools, parents, curriculum developers, the business community, inspectorial services, examination bodies, local (e.g., state) decision makers, national decision makers, and politicians (see Passey & Ridgway, 1991) operating at different levels of social organization. Each is governed by a unique set of constraints. Each subcommunity has its own unique pattern of demands upon, and rewards for, people working within. World views about the nature of education are often different. So are time tables for action, the nature of the tasks being performed, and the quality and quantity of evidence which is taken into account when decisions are made. Communication between people working within these subcommunities is far from perfect. A degree of insularity exists, and mutual perceptions are not always positive.

Assessment practices reflect perceived needs for data gathering which might inform educational decision making at both macro and micro levels. The interconnectivity of education systems is likely to increase; an awareness of the range of ambitions of different people, and their beliefs about the nature of education and its assessment is likely to become increasingly important. Sample ambitions of different people within different subcommunities are shown below:

- *Student*
  "I need the grades" (implicit "exchange" or "economic" view of learning)
  "I can see some progress" (implicit "personal development" view of learning)
  "I managed to get the grade!" (implicit "hurdle" view of achievement)

- *Teacher*
  "I can plan lessons when I see progress" (implicit "ladder" view of mathematics)
  "Diagnosis lets me teach with surgical accuracy" (implicit "constructivist" view of learning)

- *Department*
  "We need to maintain standards" (implicit "social status" view of mathematics)
  "Mathematics is a discipline" (implicit "process" view of mathematics)

At each level one needs to ask: What are the ambitions of the people at this level? What range of beliefs do they hold about assessment? How does education fit into their system of values? Individuals within particular levels make different uses of assessment. Some examples are given below.

Societal Level; assessment can:

- act as a fine grained filter to allow access for the most able to the upper reaches of society;

- defuse social unrest by allowing access to greater privilege to the most able of the underclass;
- have the role of explaining away social inequality in terms of academic merit.

Industrial Level; assessment can:
- certify competence for professional groups (and can define minimum standards of conduct which customers can expect);
- legitimize the selection processes of employers (which often have no validity, and maintain patterns of social inequality).

Family Level; assessment can:
- confer or reduce status;
- enhance social partitioning;
- lead to disgrace, even suicide.

Student Level; assessment can:
- enable self-actualization as a mathematician;
- be proof of acquiring a range of general purpose thinking tools;
- be used to gain access to or be denied a particular career or further education (with direct impact upon income, standard of living);
- show it is possible to meet set criteria, such as attainment targets;
- confer access to a desirable spouse or social class;
- inhibit the concept of learning for its own sake;
- emphasize a utilitarian (or exchange) view of knowledge;
- reduce self esteem.

Teacher Level; assessment can:
- be seen to impart self esteem or public esteem (in terms of the value systems current in education);
- provide evidence for promotion;
- demonstrate minimal competency;
- be claimed to help and support students;
- claim to support self-improvement;
- restrict teaching to the examination syllabus.

Departmental Level; assessment can:
- ensure adequate examination passes to maintain power and status within the school;
- satisfy student entitlements.

School Level; assessment can:
- gain sufficient physical and human resources by attracting well-qualified pupils;

- increase or decrease the status of the school within the local community.

Support Service Level; assessment can:
- identify schools and individuals in need of assistance;
- identify mathematical problem areas.

Curriculum Development Level; assessment can:
- help teachers assess the efficacy of new materials;
- define what is to be learned.

Inspectorial Service Level; assessment can:
- be claimed to indicate and maintain "standards";
- be used to monitor student performance.

Education Department/Office Level; assessment can:
- be used as a guide to create policy to support school improvement;
- be used to indicate effectiveness in terms acceptable by the government in power;
- be used to indicate efficient spending of public money.

Political Level; assessment can:
- be used to establish a positive public profile for the government by being seen to be concerned;
- be used to claim value for public money;
- be used to impose a particular world view upon society by selective emphasis in the school curriculum (because it is "right", and to increase the chances of re-election);
- be used to legitimize social inequalities in salary and status;
- be used to offer a rationale for economic failure.

International Level; assessment can:
- if internationally recognized, increase the mobility of the workforce to the direct benefit of countries which offer the largest salaries;
- undermine the capacity of developing nations to improve the quality of teaching (or industry), because of the outflow of skilled persons.

Assessment serves different needs for a large number of subcommunities. Changes in assessment practice will depend upon the extent to which the needs of the most powerful subcommunities, in terms of driving or resisting innovation are served by the proposed changes. These will need to be addressed before any general revision of an assessment system can begin.

4. THE PSYCHOLOGY OF ASSESSMENT

A Focusing Question: *What Is It That We Want Students To Be Able To Do*?
The first part of this chapter offered a broad conceptualization of the roles which assessment plays in society at large. This overview is important if mathematical educators are to address the practical problems of how to bring about a significant revision of current systems of assessment. This section deals directly with the psychology, as opposed to the sociology, of assessment.

Some Assertions

1. The choice of measures of anything should reflect one's theory of what the phenomena of interest are.

2. The ways that student performance is assessed, if the assessment has any importance, can drive classroom activities. It is important that classrooms are driven to places that those concerned, e.g., mathematical educators, actually want them to go to. It follows that educational ambitions should define assessment systems.

3. Monitoring, evaluating and improving student learning is not simply a matter of finding a few nice techniques:
   - fundamental questions have to be asked about long term ambitions and how they can be achieved;
   - assessment should be a central issue in curriculum design and implementation, not something to be bolted on afterwards;
   - time spent on assessment should be seen as an important component of teaching time;
   - the costs of assessment need to be justified.

4. Assessment schemes need to do more than simply say what is assessed. Methods for the analysis of information, and the ways that different decisions flow on from different results, need to be spelled out.

5. There are two distinct uses for assessment, which should not be confused. The first is for self-assessment, review, and change. The second is to demonstrate some levels of achievement and to satisfy demands for public accountability.

6. An early focus on the statement of clear aims and objectives can be a fast road to the trivialization of the educational process:
   - one can overemphasize things that are easy to measure;
   - educational aims are quite diffuse, and should not be hidden, e.g., many people want students to develop: critical and creative thinking;

positive feelings, values, and attitudes; higher-order cognitive skills; negotiation, expressive and communication skills; and process skills such as the ability to conduct investigations;

- aims and objectives mean almost nothing without concrete examples of the sorts of things that students will be able to do at the end of the course, and beyond.

7. The issues of reliability and validity of different forms of assessment need to be explored; actually, the distinction between the two is easy to make in very simple cases, but falls apart in cases like the assessment of mathematical knowledge. At least as important is exploring their corruptibility. Measures can be reliable and valid when they are used for research purposes to predict some desirable outcome, then become useless when they are used as performance indicators themselves. For example, broadly based psychometric tests of mathematical attainment may be good predictors of the mathematical competencies of children when used for research purposes, but would be invalidated completely if teachers "taught to the test".

8. Assessment schemes are embedded in educational systems which operate in broad social and political contexts. They often reflect social and political values, and implicit beliefs. It follows that assessment schemes may prove to be very resistant to change.

Some New Concepts in Assessment

There is a considerable literature about assessment in both psychology and education. Anyone with roots in these traditions will want to know about the reliability of different measures, e.g., How similar would results be if it were possible to take the same measurement twice?, and about the validity of these measures, i.e., What are they measures of? What are the dimensions of the phenomenon of interest? A major problem with much of this work is that it is overly concerned with technical issues such as reliability, validity, item analysis, scaling, factor analysis, the search for an understanding of the structure of particular domains, ways to increase measurement accuracy, and the like, and it is too little concerned with conceptual issues about the uses to which tests are put, and the relationship between testing and teaching. As all good constructivists know, there is little point in offering someone a rival intellectual framework unless the person is dissatisfied with their current position. To overcome the apparent *impasse*, some useful additions to the psychometrician's collection of intellectual tools will be offered.

Ideological validity. Assessment schemes reflect beliefs about what is important in any activity. Ideological validity refers to the educational, moral, philosophical and political values that are implied by use of any particular assessment scheme.

Generative validity will refer to the changes in behavior which occur because a particular set of measures is used, (e.g., using tests of arithmetic to assess a school's relative standing in mathematics is likely to drive the curriculum towards basic skills, and away from conceptual understanding). Generative validity is knowable in principle, but can only be determined after a particular set of measures has been adopted.

Tentative generative validity. Decisions about what measures to adopt should be governed by judgements about the likely generative validity of different measures, and one's personal judgements of their long term value to the educational process. Tentative generative validity (TGV) identifies likely directions of change and their inherent value. It follows that a measure can have as many TGVs as there are raters.

The corruption coefficient. The corruption coefficient (CC) measures the extent to which scores can be raised on a particular measure without changing the phenomena which the measures are supposed to relate to, e.g., coursework scores can be improved by teacher input, without benefiting the student's understanding.

Stick-and-carrot validity. Stick-and-carrot validity assesses the extent to which an assessment system can be used to control the education system. If student learning is to be enhanced, one must first achieve a consensus on appropriate ideological validity, then ensure that measures are chosen with a high TGV and a low CC. It is also necessary to ensure that stick-and-carrot validity, when it is high, is used for beneficial rather than negative purposes.

5. METHODS TO LOOK AT THE ASSESSMENT OF MATHEMATICS IN ANY COUNTRY

The aim of this section is to sketch some approaches which might be taken to understand the role of assessment within the educational system of a particular country. It will establish some criteria for judging assessment schemes, draw attention to some of the agents who are important in determining educational policy in action, and then to review the assessment system which is in place. Because of the differences between countries in the ways that education is organized and its place in the broad social context, all that can be offered is a set of starting points for analysis. One outcome of such an analysis would be a better understanding of the roles played by assessment at different levels within the education system and the society. The analysis could provide a platform for change. The three activities below have been selected from a larger set which appears in Passey & Ridgway (1991).

Activity 1: Criteria for Judging Assessment Schemes
The aim of this activity is to set out some criteria against which any particular assessment scheme can be judged. The list below is offered as a starting point for discussion. You might like to decide which criteria are relevant to your views of assessment by adding to and deleting from the items below. A more difficult challenge is to satisfy these desirable features.

A scheme for the assessment of students should:

- evaluate things which are seen to be of value;
- ensure that everyone with an interest or concern has a say;
- be understood by everyone with an interest or concern;
- demonstrate and build on current successes for students, teachers, and schools;
- collect information about things which can be acted upon;
- use measures or forms of evidence of adequate reliability;
- consider ways to monitor individual, course, and school progress and performance;
- relate to the mathematical development of pupils;
- address teacher development at the level of the individual, the team, and the teaching community in general;
- have a clear role for both internal and external assessment;
- define course aims and objectives by providing exemplars of the tasks which students will be able to perform at the end of the course.

A scheme for the assessment of students should also:

- enshrine (and perhaps clarify) the value systems of the teachers and disciplines which it serves;
- be designed so that carrying out the assessment tasks should itself be of educational value;
- be designed so that the costs of establishing and administering the assessment system can be easily justified in terms of the benefits to students over and above the benefits they would have received had the same resources simply been devoted to teaching;
- have systems to review the whole assessment process.

Activity 2: Assessment by Layers
The second activity sets out to identify the key agents within the educational system.

- Starting with the list of subcommunities offered earlier, you might wish to identify the key players in the education system of interest. Add to the list, or delete subcommunities.
- What assessment is appropriate? What is assessed? Who does the assessment? For whom? What happens as a result of the assessment? How does information flow between subcommunities?

Next assess the dominant views of assessment implicit in these uses (again using some of the earlier ideas).

- What functions does assessment serve in each subcommunity? What values about mathematics are implicit in the ways that mathematical assessment is used by each subcommunity?
- How much power does each subcommunity have to: Control educational processes? Initiate widespread change? Inhibit change? Change the assessment system?
- Can you discover ways to work with powerful groups to encourage a style of assessment which you judge to be more desirable than the one currently in use?

Activity 3: Analysis of the Current Assessment Profile
Review both the column and the row titles in the table below, and amend them appropriately. Choose some assessment which is significant in some way, for example, which is used to gain access to employment or higher education. Then make some estimates of the percentage of the final mark which is derived from each component. Obvious questions relate to the balance of activities and to the ways that the balance could be improved. More difficult questions relate to ways to set about changing current practices.

Assessment Method	Assessor		
	Student	Teacher	External exminer
Timed written examinations			
Multiple-choice tests			
Extended tasks			
Practical work (construction, experiment...)			
Statistical or methodological work			
Process skills			
Portfolios of work			
Class tests			
Oral representations			
Open-ended problems			

Table 1 *The assessment profile*

The pattern that results can be used to reflect on the messages about the nature of mathematics being conveyed to students, teachers, parents and society.

It is appropriate to reflect on the ideological validity, generative validity and the corruption coefficient of this assessment profile. The tentative generative validity of any proposed changes need to be discussed before stick-and-carrot validity is explored empirically.

6. CONCLUSIONS

Assessment is a murky business. It raises fundamental issues about our beliefs on the nature of mathematics and knowledge in general, teaching and the educational process, and the relationships between the individual, school, and society. Awareness of the plurality of views is an important starting point for reform. Attempts to use improved assessment practices to drive desirable educational change are laudable, but likely to be difficult to achieve in practice.

REFERENCES

Cameron, J., Cowan, R., Holmes, B., Hurst, P. & McLean, M.: 1983, *International Handbook of Educational Systems (Vols. 1–3)*, Wiley, London.

Department of Education and Science: 1989, *Mathematics in the national curriculum*, Her Majesty's Stationery Office, London.

Greer, B. & Mulhearn, G.: 1989, *New directions in mathematics education*, Routledge, London.

Husen, T.: 1967, *International study of achievement in mathematics: A comparison of twelve countries*, Almqvist and Wiksell, Stockholm.

Little, A.: 1990, 'The role of assessment, re-examined in international context', in Broadfoot, P., Murphy, R., and Torrance, H. (Eds.), *Changing educational assessment: International perspectives and trends*, Routledge, London.

Mathews, J.: 1985, *Examinations: A Commentary*, George Allen and Unwin, London .

Passey, D. & Ridgway, J.: 1991, *Building on success-evaluation and self-evaluation for advisory teachers*, National Council for Educational Technology, Coventry.

Ridgway, J.: 1988, *Assessing mathematical attainment* NFER-Nelson, Windsor.

Robitaille, D.F. & Garden, R.A. (Eds.): 1989, *The IEA study of mathematics II: Contexts and outcomes of school Mathematics*, Pergamon Press, Oxford.

Shimahara, N.K.: 1985, 'Japanese education and its implications for US education', *Phi Delta Kappan*, February, 418–421.

Stigler, J.W., Lee, S-Y., Lucker, G.W. & Stevenson, H.W.: 1982, 'Curriculum and achievement in mathematics: A study of elementary school children in Japan, Taiwan and the United States', *Journal of Educational Psychology*, **74**, 315–322.

Jim Ridgway & Don Passey
Department of Psychology
University of Lancaster
England

PETER GALBRAITH

PARADIGMS, PROBLEMS AND ASSESSMENT: SOME IDEOLOGICAL IMPLICATIONS

"So my answer to the question "How do you know? What is the source or basis of your assertion? What observations have led you to it?" would be: "I do not know: my assertion was merely a guess. Never mind the source ... if you are interested in the problem which I tried to solve by my tentative assertion, you may help me by criticizing it as severely as you can" ..." (Karl Popper, 1968)

1. PARADIGMS

The current ferment with respect to mathematics assessment involves both substantive issues, i.e., how to assess different components of mathematical activity; and wider issues, i.e., the role and purpose of mathematics assessment within society at large. It is essential in such times to step back in order to evaluate the nature of the wood, as well as attending to the individual trees.

This chapter locates issues of mathematics assessment in a wider educational and societal context and to this end it is appropriate to review the major *paradigms* that guide research and inquiry in the social sciences — specifically education; see for example Phillips (1987). For simplicity, a paradigm in the Kuhnian sense is regarded as a basic set of beliefs or assumptions that act as reference points in guiding and directing our activity (Kuhn, 1962). Guba & Lincoln (1989) provide extensive discussions of what they term the conventional (or scientific) paradigm and the constructivist (or interpretive) paradigm. Carr & Kemmis (1986) add a third dimension in elaborating a methodology based on a paradigm derived from critical theory, (Habermas, 1971), which is elaborated with respect to education in works such as Gibson (1986) and in relation to mathematics education by Skovsmose (1985).

Conventional and Non-conventional Paradigms

Conventional Paradigm

This paradigm, commonly referred to as *the scientific paradigm* of inquiry is characterized by the following assumptions (selected).

M. Niss (ed.), Investigations into Assessment in Mathematics Education, 73–86.

(a) There exists a single reality that is realized in universal natural laws that can be verified by objective observation.
(b) Facts and values are independent. Facts can be uncovered independently of the values of an observer or values that may be invoked to interpret them.
(c) Scientifically identified phenomena have widespread significance and problem solutions widespread application across contexts and over time as generalizations.
(d) By eliminating or controlling for confounding influences inquiry can converge upon truth and explain, predict, or control nature as it really is.
(e) Scientific inquiry is the natural way to determine the definitive and enduring truth about actual states of affairs.

Constructivist Paradigm

(a) There are multiple socially constructed realities and truth is the best informed construction about which there is presently consensus.
(b) Facts and values are interdependent. Facts have no meaning except within some value framework.
(c) Problem solutions have local applicability only, and when introduced into new contexts will be at least as much affected by these contexts as will affect them.
(d) Inquiry is problematic — consisting of humanly devised ways to entertain constructions about situations that are subject to continuous refinement, revision, and when necessary replacement.
(e) Phenomena including problems exist only within some construction, and have no meaning except in the context for which the construction occurred.

These assumptions are most readily comprehended by contrasting them with the respective counterparts given above.

The constructivist paradigm has risen to prominence as workers in the human sciences have become increasingly skeptical of the usefulness of the conventional approach in many situations. The initial widespread use of the conventional paradigm in social science inquiry may be attributed to the successes claimed for the approach in advancing knowledge in the natural sciences. Below, some illustrations are provided pertaining to the interdependence of facts and values.

A baseball pitch is neither a ball nor a strike until evaluated by an umpire whose interpretive judgement decides what will become a "fact" of the game. The "fact" is inseparable from the evaluative interpretation that brought it into being.

As a further example, imagine a group of individuals making observations to determine the value of the acceleration due to gravity at a given location. All arrive at a value of 9.8 ms^{-2}. The conventional view would see this in terms of the independence of scientific facts from the values of observers. The constructivist would note that a unanimous consensus had been achieved — the situation is unproblematic. The scientific outcome is the same but the way in which its meaning is interpreted in terms of inquiry is quite different.

The constructivist is prepared to examine a set of results and to consider their possible application to other situations given the contextual features of both. He/she is not happy with the notion of context free generalization. Transferability, rather than generalizability characterizes this aspect of the search for consensus.

It is worth noting that constructivism, as it has emerged within the literature of teaching and learning, for example with von Glasersfeld (1987), and Wheatley (1991), has its roots in this wider field of contemporary philosophical activity. That is, the concepts and structures found in constructivist literature in education have their roots somewhere in the assumptions listed above. Within mathematics education the emphasis is on alternative constructions of meaning by individuals and the implications of such for learning and assessment.

In a subject such as mathematics education in relation to the assessment of mathematical abilities we must answer the following questions:

(1) What is there to know? — The *ontological question*
(2) How is the known constructed? — The *epistemological question*
(3) How do we know when something (about mathematics) is known? — The *methodological question*
(4) How do we go about finding out what is known? — The *implementation question*

Question (1) has to do with the substance of mathematics, and (2) is concerned with how knowledge is constructed and transmitted. Questions (3) and (4) pertain specifically to assessment although (1) and (2) are also relevant.

For some aspects of mathematics, there is uniform consensus on what constitutes correctness, e.g., for algebraic material derived by processes of logic from underlying axioms. Other aspects are, however, problematical, e.g., there is no present consensus on what the substantive content of problem solving entails. So if there is no uniform agreement across the substantive areas of mathematics itself the situation is immeasurably compounded when we incorporate the human and social elements of the education component — specifically in this case assessment. The way in which we develop assessment procedures in relation to questions (3) and (4) above depends crucially on our underlying belief systems about mathematics and what is involved in "knowing" mathematics. In that regard, the conventional and constructivist paradigms tell us some very different things.

Critical Paradigm[1]

The third paradigm, derived from *critical theory* (Habermas, 1971; Gibson, 1986; Carr & Kemmis, 1986), is not so easily described by assumptions but rather in terms of the constructivist approach with which it shares some sympathy. The critical paradigm agrees with the constructivist insistence that reality is constructed, and rejects the claims of sanitized objectivity of the conventional approach. However, critical theorists believe that those working within an interpretive framework are too passive in that the framework itself is not critically examined for distortion and bias, i.e., crucial problems of conflict and change, are passed by through the acceptance of existing reference points.

Essentially, as summarized by Carr & Kemmis (1986), adherents to the critical paradigm are concerned that frameworks used to interpret the world are subject to illusory beliefs and irrational influences, i.e., they are distorted. Using a purely interpretive approach in such situations means that an emancipatory dimension is lacking — a dimension that seeks not merely to observe and interpret but to transform as a means for identifying distortions in the self-understanding of practitioners and for surmounting them. It has a sociopolitical arm.

The perceived tension between mathematics education as it is habitually practiced and the emancipatory ideals of critical education is raised in a paper by Skovsmose (1985). In lamenting that mathematics education and critical education scarcely touch, he infers that the former (in particular structuralism) provides a potent means for "socializing students into the technological society" — of which the examination system is an entrenched artifact.

The implied acceptance of existing conditions that critical educators would dispute is illustrated in Howson & Mellin-Olsen (1986, p. 27) where a section heading reads "Adjusting Teaching Strategies to Pupil Expectations and Examination Demands". The critical perspective would argue that pupil expectations are coerced by the examination system. The authors refer to works such as Bourdieu & Passeron (1977) and to concepts of cultural capital and symbolic violence in relation to inequality of opportunity to succeed within the system. While the plea is to be sensitive to implications of inequalities deriving from sociological conditions, at the end of the day the common examination is accepted as the lot of all. The critical approach would instead address the integrity of the examining and certificating system as such, with a view to transforming it.

Educational theory must, from the critical perspective, offer theoretical perspectives which empower educators with the means for reducing or eliminating aspects of the existing order which are invalid or unjust.

In a paper focused towards issues of assessment Bates (1984) takes a critical perspective in which he identifies perceived conflict between evaluation conducted for *managerial* versus *educational* purposes and the resulting implications for schools and teachers. The purpose is to expose distortions in order to better equip administrators and teachers to improve educational practice. Foucault (1977), in discussing the relationship between knowledge and power, refers to threats to intellectual elites that emerge when the populace no longer need them as pipelines to knowledge. It is then, he argues, that systems of power block, prohibit, and censor wherein intellectuals as agents of the system assume responsibility for knowledge brokering and discourse through legitimation procedures erected through validation schemes such as examinations. This amounts to the controlling of what counts as credible knowledge and hence the accreditation agenda by the intellectual powers that be. There are those who claim that the gatekeeping role of mathematics makes its perpetrators among the most guilty in this respect.

There will be a body of educators who believe that matters of educational practice, including assessment, are academic and professional issues that should remain outside politics. The response of critical educators would be that practices involving imposed constraints and discriminatory activities are inherently political and should be recognized and treated as such. To ignore them is itself a political act.

A number of contemporary assessment issues will be addressed here from the perspective of paradigms. The discussion will center upon the educational implications following from different paradigmatic stances and not upon the justification of any particular stance.

2. PROBLEMS OF ASSESSMENT

Generalized Achievement Testing

Firstly, support for statewide, national, or international testing rests upon conventional assumptions that some generalizable mathematical truths usefully will emerge and can be validly measured and compared across contexts. Constructivists and critical theorists would have nothing to do with such procedures. Constructivists would regard the process as invalid since mathematics in any place is viewed as contextually constructed. This view has gained increasing support with the emergence of culture-based mathematical orientations to curricula (Bishop, 1988; D'Ambrosio, 1985; Stigler & Baranes, 1988; Wilder, 1981). For example, Wilder (1981) stresses that each culture has its own mathematics that evolves, lives, and dies with the culture, and that school children also bring to school a mathematics of their own. Stigler & Baranes (1988) write that:

> "a new sociology of mathematics has arisen that takes as its premise that the foundations of mathematics are to be found through examination of the cultural practices in which the activities of mathematicians are embedded." (p. 253)

Adherents of the *critical paradigm* would go further than rejection of national and international testing on the basis of validity to note also the managerial purposes that such testing serves. For example, cross-nation or statewide comparisons provide data that governments and multinational interests can use in manipulating the education system. D'Ambrosio (1985) writes of the damage induced in some developing countries by excessive authority reinforced by "culturally absurd evaluation mechanisms" that kill natural components of insight, creativity, and ingenuity. He cites Luna (1982) who, in writing from the context of the second IEA study, argues that in some countries at least, mathematics as a school subject contributes to the demise of insightful numeracy through a search for mastery of formal "basics" that is unrelated to, and even inhibits, intelligent functioning in new situations. An inference is that pressures on school systems applied through national responses to IEA data actually degrade mathematical performance in such countries.

It should be noted that comparative studies of mathematics performance such as described and advocated by Stigler & Baranes (1988) belong to a different genre. Testing associated with project-based studies of mathematical learning is a necessary component in understanding how mathematics learning differs and develops in cross-cultural contexts. The purpose is strictly educational.

The huge sums of money spent on national and international testing programs, should, from the critical perspective be redirected to improving the quality of resource support for mathematics teaching in those countries, regions, or schools that are most in need. A question for the mathematics education community is that of values conflict. Is it consistent to embrace a *constructivist* approach to teaching but then support assessment programs which deny the constructivist stance? Is it consistent to argue for school-based assessment on the one hand and to support statewide, national, or international competency testing on the other? Does it matter what is being tested?

School Policy and Assessment

A related version of the same issue occurs at local levels within school mathematics departments where there remain a substantial number of mathematics teachers, including some departmental heads, who believe that formal, preferably external, testing remains the only valid form of mathematics assessment. Frequently this situation is reinforced by school assessment policy and is a potent adversary to curriculum reform. Debates at the level of assessment as such are often not useful as they do not expose the underlying belief structures. Conflict in regard to what constitutes valid mathematical knowledge and consistent ways of assessing it lies at the root of the matter. A first step to improving practice is to begin by addressing the problem at the basic level of belief systems.

A similar situation arises when there is inconsistency between approaches to learning and approaches to assessment. A concise discussion of the *constructivist* point of view in teaching is given in Schoenfeld (1987, pp. 20–30). He writes that adopting the constructivist perspective allows us to see what happens in classrooms quite differently:

"the key issue is how students will interpret what you show them rather than how much of it they will absorb. ... your job begins rather than ends with a clear presentation of material." (pp. 20–30)

Increasingly this approach to learning is being accepted in schools. Yet when it comes to assessment, the students are no longer allowed their contextual interpretations but are expected to demonstrate abstracted generalized knowledge in answers to standard test questions. This is mixing metaphors with a vengeance. Again the problem is a lack of consistency due to a lack of attention to the underlying paradigms supporting the respective activities of teaching and assessing. The move to contextualize assessment within the day to day processes of teaching represents substantial progress towards removing inconsistencies. A further implication is that exciting material, such as contained in the *Australian Mathematics Curriculum Teaching Program* (Clarke, 1988), should be viewed as a whole with assessment procedures and issues integrated closely within the curriculum development. That is, it is inappropriate to look at alternative modes of assessment without also asking questions about associated teaching approaches and assumptions underlying both.

Formal Versus Constructed Knowledge and Performance Assessment

A fourth issue has to do with concepts of valid mathematical knowledge and is exemplified in contrasting ways within the work of Collis & Romberg (1990) and D'Ambrosio (1985). The former paper is set within the purposes of the 1989 *Standards* document of the National Council of Teachers of Mathematics (NCTM), U.S.A., and addresses the challenge to design test material to validly measure achievement in areas such as problem solving, communication, and reasoning. While using terms such as empowerment (a term common among critical educators) the paper appears to remain firmly within the traditions of the *conventional* paradigm with its concerns of validity, reliability, etc., of test items that will be applicable to a large population of students. One particularly interesting section of the paper reviews research that reported on mathematics used by workers in their job situations in relation to the symbolic structures of school mathematics. The following contrasts were reported.

Work Place Mathematics	*School Mathematics*
o Done in social context with others presents	o De-personalized
o Procedures adjusted to fit adjustment situation	o Formalized rules — no adjustment
o Social, cultural and emotional factors involved	o Such factors not involved
o Motivation related to task-in-hand goal	o Motivation related to distant goal
o Needed data not defined and need to be selected	o Needed data well defined
o Tasks are particular, singular and individual	o Tasks aim at deriving generalizations
o Precision required is defined by the situation	o Results expected to be precise and unambiguous

When these attributes are viewed in terms of underlying assumptions, it is noted that *Work Place Mathematics* represents an embodiment of all that is associated with the *constructivist paradigm*, while *School Mathematics* exemplifies the *conventional paradigm* to a similar degree.

The paper then proceeds to discuss the limitations of Work Place Mathematics in terms of context specificity and non-generalizability across differing empirical settings — that is, a value position is stated with regard to the paradigms. The goal is to develop procedures to raise the level of functioning from workplace methods to use of generalized symbolic forms and to devise test items that will measure creative and problem solving capabilities in a generalized and stable way. Clearly the conventional paradigm is driving the work here.

Constructivists on the other hand would argue that the concept of generalizability is irrelevant as presented. Given a new work situation effective strategies would again be constructed and such would occur for each succeeding change in job demand. It should be noted that some of the "homegrown" mathematical strategies used in the workplace were quite sophisticated — they just were not standard generalized school mathematics procedures. The examples highlight differing views as to what are appropriate ways to assess, for example, problem solving ability. Does success on de-contextualized assessment tasks originally developed from observed problem solving behavior give a valid measure of a person's ability to handle a previously unencountered problem? Is such success an adequate measure of the capacity to construct a solution to a new challenge? Yes! say the conventionalists; No! say the contructivists.

It is fascinating to note that D'Ambrosio (1985) precisely reverses the priority between work place and school mathematics. He notes that *ethnomathematics* is practiced by uneducated people, sometimes by virtual illiterates, but adds:

"The learned matheracy eliminates the so called "spontaneous" matheracy. An individual who manages perfectly well numbers, operations, geometric forms and notions, when facing a completely new and formal approach to the same facts and needs creates a psychological blockage which grows as a barrier between the different modes of numerical and geometrical thought ... Once going to school the tendency is to lose these capabilities and not be able to replace them by "learned" ones ..." (p. 45)

He goes on to relate that social communication in dealing with economics, prices, building, urban planning, and administration is rendered much more difficult. This effectively says that problem solving and modeling capacities are actually eroded when methods based on contextualized constructed realities are replaced by approaches based upon the de-contextualized, generalized, abstract procedures of formal mathematics that find their educational justification within the belief framework of the conventional paradigm.

The question is not whether the development of items such as Collis & Romberg are seeking is worthwhile. There seems little doubt that they have the potential to provide valuable information about aspects of student performance in areas that have been lacking. Indeed Stigler & Baranes (1988, p. 259) specifically address the balance between "universals" and "specifics":

"Human cognition undoubtedly contains universals, both in the forms of mechanisms of functioning and in the form of specific knowledge structures. Our argument is not that these universals are unimportant, but rather that non-universal aspects of cognition, and in particular those that are culturally specific, are of equal importance." (p. 259)

The question is what de-contextualized tests of process collectively measure and where and for what purpose they should be used? Certainly constructivists would dispute that they could, in themselves, test the aims of a new syllabus.

It could be noted that critical educators would take an additional stand noting how much the development of items appears to be motivated by the desire to produce valid, widely applicable test material. They would want to peruse the impact on proposed changes and goals of pressures induced by the presence of a multimillion dollar test industry and its traditional impact on school assessment.

This discussion leads naturally to a consideration of approaches to the assessment of problem solving, investigation, and mathematical modeling. In some countries and states, the freedom to innovate is controlled by external assessment requirements. The question of what is actually being measured by the inclusion of open-ended problem type questions in traditional test situations is central. The constructivist position is that whatever the collective results of such a test might mean they cannot be used as a formal index, e.g., of problem solving ability, as problem solving involves the construction of solutions to problems in context. The fallacy of the approach from a constructivist view is in separating out and making explicit part of the world, and then re-introducing it as the whole world. It is as though judges at an Olympic skating event decided that comparison of competitors in a free routine section was impossible except by some common formula involving the separation of certain skills identified as common components of all routines. We know full well that performances of Torvills and Deans far transcend that kind of analysis and their excellence is demonstrated idiosyncratically in the completeness and consistency of their performances.

3. A POLITICAL INFERENCE

In general, those with a *critical* perspective, while welcoming the partial inclusion of school-based assessment and greater variety in external assessment requirements, would regard these as incomplete measures, acceptable in the short-term, but not as long-term solutions. The concept of using external assessment requirements to influence teaching, while perhaps defensible in the short-term, would be seen as ultimately disempowering to teachers in impeding the growth of full professional responsibility, and to students in making their choices and interests irrelevant. Such a view has both professional and political consequences.

Professionally it becomes essential to address the question of comparability of standards across teachers, schools, and districts through processes of moderated peer review. The teacher judgement literature, together with the concept of explicitly stated standards supplemented with exemplars, provides a basis for progress. *Constructivist* views that problem solving, modeling, and investigation work must be evaluated developmentally and holistically in context adds the mathematical dimension to the case. The political dimension arises when a soundly-based educational case is used to exert continuous pressure on authorities to amend official requirements and practices. A critical stance would demand nothing less. This view specifically refuses to accept externally imposed assessment methods as permanent defensible constraints.

4. FINALE

Perhaps it is appropriate to conclude by summarizing the main thrust of this paper as a series of issues/questions to be addressed.

(1) *The need to confront inherent contradictions that exist when constructivism drives curriculum design and knowledge construction, but positivistic remnants of the conventional paradigm drive the assessment process.*

The pseudo support embedded in expressions such as "I like the way you have developed your mathematical approach, but we'll have to wait for the common exam results to see if it's been successful" is a typical illustration of paradigm confusion. If, as argued by Popper (1968) and reinforced by Phillips (1987), a theory can only be tested in terms of its own tenets, then the success of a constructive approach to learning can only be evaluated by assessment procedures derived from the same principles. Context independent generalized testing is not only unjust to such a cause; it is invalid, and the invalidity is in no way diminished by including socalled *tests of process* in the common instruments.

(2) *The attitude of the mathematics education community towards national and international achievement testing that is used for cross-cultural and cross-group comparisons.*

The very real difference in purpose that exists when results are used managerially (as in the political use of gross national comparisons based on notions of context free generalization) and educationally (as in searching for insights about how mathematical understanding is constructed across cultures and groups) reflects a deep division between corresponding belief systems. The diversion of the large sums of money involved to improve the lot of mathematical literacy in identified areas of need would be a vigorously pursued goal if the critical paradigm was more vigorously propounded within the mathematical education community.

(3) *The nature and purpose of external assessment in its various manifestations, and the traditional acceptance of such procedures as superior to developmental school-based procedures that make consistent use of informed teacher judgement.*

There is a challenge presented in the emerging literature on teacher judgement and standards-referenced assessment Hoge & Colardarci (1989), Sadler (1987), particularly with respect to the assessment of activities such as problem solving, modeling and investigation work in which class teaching employs constructivist principles. The replacement of cross-sectional additive assessment tasks by cumulative longitudinal performance, data aimed at the identification of stable performance levels, places increased stress on teacher knowledge, professionalism, and concepts of comparability.

(4) *A critical evaluation of the societal purposes served by examination systems, and the tensions experienced when a profession historically devoted to reproducing itself faces the criticism that in so doing it is safeguarding the privileges of its members and also reproducing the inequalities in society.*

For example does (should!) the mathematical education community accept existing restraints in the form of examination and certification systems, or does (should!) it work to transform or remove them according to higher principles of validity and equity? To what extent are the real issues clouded by underlying beliefs which are instrumental in maintaining inadequate, invalid, or unjust practices and which prevent thorough going reviews of what is meant by terms such as "standards"?

(5) The tension between the values ascribed to school mathematics and to workplace or ethnomathematics in whatever society or culture they occur.

Contradictions are implied when tests of process couched in terms of conventionally-based school mathematics purport to give valid measures of ability with respect to culturally embedded applications, and when concepts of generalizability are defined for conventional mathematics, contest with concepts of transferability as needed within workplace or ethnomathematics.

(6) Mathematics as a means of socialization into a technocratic society rather than as a critic of it.

There is a challenge to mathematics educators to share the concern of the critical paradigm for an examination of the social role of mathematics with respect to reproduction and privilege, and for the inclusion of social critique in relation to mathematics and its assessment, if not in school, then at least within tertiary education and, in particular, teacher education.

So ultimately the issue of assessment must address the implications of underlying belief systems. It is not sufficient to take an instrumental or "bag of tricks" approach to methods of evaluating student learning or be satisfied merely to work within whatever official constraints apply at the time. There is much enthusiasm and agreement about modern directions, about what is worth knowing in relation to mathematics as for example in the NCTM *Teaching Standards* (1991). But developments in teaching approaches and assessment need to be related more critically and consistently to the implications of underlying assumptions than has been the case to date. It is a challenge we all share.

NOTE

1. The term *Critical Theory* is, strictly speaking, associated with output from the Frankfurt school. The term *Critical Paradigm* is used more generally to capture elements of theory and practice, derived and practiced by a wider group of scholars whose work is descended from, or in articulated sympathy with, the founders of the school.

REFERENCES

Bates, R.J.: 1984, 'Educational versus managerial evaluation', in Broadfoot, P. (Ed.), *Selection, certification and control: Social issues in educational assessment*, 127–143, Falmer Press, New York.

Bishop, A.J.: 1988, *Mathematical enculturation*, Kluwer Academic Publishers, Dordrecht.

Bourdieu, P. & Passeron, J.C.: 1977, *Reproduction in education, society and culture*, Sage Publications, New Gary Park, CA.

Carr, W. & Kemmis, S.: 1986, *Becoming critical: Education, knowledge, and action research*, Falmer Press, New York.

Clarke, D.: 1988, *Mathematics Curriculum and Teaching Program: Professional Development Package: Assessment Alternatives in Mathematics*, Curriculum Development Center, Canberra, Australia.

Collis, K.F. & Romberg, T.A.: 1990, 'Assessment of mathematical performance: An nalysis of open-ended test items', in Wittrock, M.E. (Ed.), *Cognition and testing*, Lawrence Erlbaum, Hillsdale, NJ.

D'Ambrosio, U.: 1985, *Socio-Cultural Bases for Mathematics Education*, Unicamp, Brazil.

Foucault, M.: 1977, 'Language, counter-memory and practice', in Bouchard, D.F (Ed.), *Selected essays and interviews*, Cornell University Press, Cornell, NY.

Galbraith, P.L. & Clatworthy, N.J.: 1990, 'Beyond standard models: meeting the challenge of modelling' *Educational Studies in Mathematics*, **21**(2), 137–163.

Gibson, R.: 1986, *Critical theory and education*, Hodder & Stoughton.

von Glasersfeld, E.: 1987, 'Learning as a Constructive Activity', in C. Janvier (Ed.), *Problems of Representation in the Teaching and Learning of Mathematics*, Kluwer Academic Publishers, Dordrecht.

Guba, E.G., & Lincoln, Y.S.: 1989, *Fourth generation evaluation*, Sage Publications, New Gary Park, CA.

Habermas, J.: 1971, *Knowledge and human interests*, Beacon Press, Boston.

Hoge, R.D. & Coladarci, T.: 1989, 'Teacher-based judgements of academic achievement: A review of literature', *Review of Educational Research*, **59**(3), 297–313.

Howson, A.G. & Mellin-Olsen, S.: 1986, 'Social norms and external evaluation', in Christiansen, B., Howson, A.G., Otte, M. & Reidel, D. (Eds.), *Perspectives in mathematics education*, 1–48, D. Reidel, Dordrecht.

Kuhn, T.S.: 1962, *The structure of scientific revolutions*, University of Chicago Press, Chicago, Il.

Luna, E.: 1982, *Student Achievement in Mathematics for the Benefit of Caribbean Communities and its Reflection in the Curriculum*, IACME-UNESCO, O.A.S.

National Council of Teachers of Mathematics: 1989, *Curriculum and Evaluation Standards for School Mathematics*, Reston, VA.

National Council of Teachers of Mathematics: 1991, *Mathematics Educators Release New Teaching Standards*, Reston, VA.

Phillips, D.C.: 1987, *Philosophy, science, and social inquiry*, Pergamon Press, New York

Popper, K.: 1968, *Conjectures and refutations*, Harper and Rowe, New York.
Sadler, D.R.: 1987, *Specifying and promulgating achievement standards*, Oxford Review of Education, **13**(2), 191–209.
Schoenfeld, A.H. (Ed.): 1987, *Cognitive science and mathematics education*, Lawrence Erlbaum, Hilsdale, NJ.
Skovsmose, O.: 1985, 'Mathematical education versus critical education', *Educational Studies in Mathematics*, **16**, 337–354.
Stigler, J.W. & Baranes, R.: 1988, 'Culture and mathematics learning', *Review of Research in Education*, **15**(7), 253–306.
Wheatley, G.H.: 1991, 'Constructivist perspectives on science and mathematics learning', *Science Education*, **75**(1), 1, 9–22.
Wilder, R.L.: 1981, *Mathematics as a cultural system*, Pergamon Press, Oxford, England.

Peter Galbraith
The University of Queensland,
Brisbane,
Australia

DAVID WHEELER

EPISTEMOLOGICAL ISSUES AND CHALLENGES TO ASSESSMENT: WHAT IS MATHEMATICAL KNOWLEDGE?

1. INTRODUCTION

Why should the question in my title be asked in a book concerned with assessment? Perhaps because it seems a matter of common sense that if we want to assess the mathematical knowledge of students, we need to be able to recognize that knowledge. How can we assess what we do not know? But as soon as we ask the question in its general form we notice that there has been no consensus among the mathematicians and philosophers who have tried to answer it. Common sense also dictates, then, that if we want to assess mathematical knowledge, we cannot afford to wait until the question of its global nature is settled.

In practice we have traditions and rules-of-thumb that tell us what things count as mathematical knowledge and how to recognize them, at least at the level of elementary mathematics. There is not much surface disagreement about the knowledge content of "school mathematics" in most countries of the world. It is only when the traditions are shaken, voluntarily or involuntarily, that it becomes necessary to re-validate the traditions, to find some perspective from which they can be judged. The adoption of hand calculators in schools may not essentially change anyone's view of what constitutes mathematical knowledge, but it does have a strong local impact on the particular knowledge children need to have in order to be able to make elementary mathematical calculations. And an emphasis on problem solving or "investigations" as a component of the mathematics curriculum introduces some new distinctions in the kinds of mathematical knowledge that the classroom is expected to generate. So although it is unlikely that a study of the nature of mathematical knowledge will have a direct impact on the way mathematical knowledge is assessed, it is nevertheless the case that methods of assessment are influenced to some extent by the prevailing beliefs about what constitutes desirable mathematical knowledge.

Our question is epistemological, but I doubt if anyone these days believes that this means it has a clear unambiguous answer. Each informed individual may believe he or she knows what mathematical knowledge is, and may even suppose that this view is the only tenable one, but if we look

M. Niss (ed.), Investigations into Assessment in Mathematics Education, 87–95.

around us at all the places where people have an interest in mathematics, it seems as if different social groups can have very different epistemologies concerning this special knowledge that they trade in. Whose views about mathematical knowledge are we to take on trust? Whose epistemology will tell us what we want to know about mathematical knowledge? The mathematician's? The philosopher's?

This is a metaphorical way of talking since there is no single idealized mathematician who speaks for all mathematicians, no philosopher with whom all other philosophers agree; since this way of talking seems suggestive, let us pursue it. We may find it useful to talk of the teacher's epistemology, or the student's, or the person-on-the-street's. We may even conceive that the educational system, because its structure and the activities that go on within it, effectively puts a valuation on items and forms of knowledge, implicitly enshrines a characteristic epistemology. It is likely — indeed it is obvious — that these various epistemologies are not in agreement on all points. And what is clear is that the confluence of them all (and others that I may have forgotten) can create an extremely difficult situation for the classroom teacher who is implicitly expected to act in a way which is compatible with each of them.

Assuming this multiplicity of epistemological interests, should we regard the mathematician's epistemology as specially privileged? I think we ought to be cautious unless we know that the mathematicians we choose to consult have made some special study of the question. A working mathematician, even one who also teaches, does not need to possess a formulated epistemology of mathematics and may, when asked, just fall back on some parts of the familiar folklore about mathematics that everyone tends to draw on rather than give the question the careful and sustained attention it deserves. With Hans Freudenthal, I am inclined to think that the insights of the educator can make an important contribution to an epistemology of mathematics. Educators are substantially more likely than most mathematicians to have reflected hard on Poincaré's fundamental question: Why is it that some people do not succeed in acquiring mathematical knowledge? (Poincaré, 1913).

2. INVENTED OR DISCOVERED?

The classic conundrum about mathematical knowledge is whether it is *invented* or *discovered*. We remember that Hadamard (1945) discusses the question in his well-known study. But although the antithesis is enticing and makes an intriguing question, there are other possible ways of describing the genesis of mathematical knowledge. The philosopher John Locke argued the empiricist position, which was also given a recent push by Philip Kitcher (1983), that mathematical knowledge is essentially arrived at by the same procedures as knowledge in the natural sciences. Jean Piaget gave a psycho-genetic account that places the acquisition of mathematical knowledge in a biological, maturational perspective. In the latter theory in particular, neither invention nor discovery seem appropriate words. So it is certainly possible to say "neither" in answer to the question "Is mathematics invented or discovered?"

A more suitable answer may be "mathematics is both invented and discovered". The two poles are not mutually exclusive but complementary. If we deny the element of invention we deny that people choose the directions in which mathematics develops, and if we deny the element of discovery we fail to acknowledge that mathematics often develops in ways that were not anticipated. The nature of the mathematical experience suggests that mathematical knowledge manages to marry the arbitrary and the inevitable. The great mathematician Cantor (see Dauben, 1979) expresses both sides of the coin when he says that "numbers are a free creation of the human spirit" but also writes to his friend Dedekind, when he feels himself forced to accept the counterintuitive possibility of mapping the points of a square onto the points of a line, "*Je le vois, mais je ne le crois pas*" (June 29, 1887).

The contrast of invention with discovery suggests the opposition of two related philosophies: *constructivism* versus *platonism* — though in the light of the above remarks these should not automatically be judged irreconcilable. Once we begin playing "the competing philosophy game", we are bound to recall the familiar triad, elaborated *ad nauseam* in all the books that mention the philosophy of mathematics: *formalism*, *logicism*, and *intuitionism*. There is not much mileage to be gained from adding to the examination of these, yet it seems clear that although the argument about which is more nearly correct is almost certainly an occupation for idle minds, each philosophy makes some significant statement about mathematics.

"Certainly, in some undeniable but obscure sense, mathematics seems at the same time to be a meaningless game, a subjective construction, and a source of objective truth." (Rotman, 1988)

Formalism, usually associated with the mathematician Hilbert, portrays mathematics as "meaningless marks on paper" manipulated according to agreed rules. (Hilbert may have been a formalist in mathematics but fortunately he was an intuitionist in metamathematics, Rotman suggests). Brouwer's intuitionism pictures mathematics as essentially "a languageless activity". Instead of assigning logicism to Frege, as we could do, let us here diagnose it as a variant strain of platonism and ascribe it, with tongue in cheek, to all working mathematicians. The philosophies are so incompatible (though each, as said above, holds a truth about mathematics) that they cannot possibly be combined to give a synoptic view of mathematical knowledge.

The central difficulty of these philosophies, each of which talks of mathematics as if it is an activity that only individuals do, is that not one of them is able on its own to make it plausible that a crucial part of mathematical activity is concerned with persuading, arguing, proving: processes that can only be carried on when other people are involved, in social groups, in communities which own a common language. Among other things, mathematics is a very special form of human discourse.

Rotman, in the paper from which the above quotation was extracted, suggests that a dialectic interaction between the activities enshrined in the formalist and intuitionist philosophies gets closer to describing the nature of mathematical activity than either does on its own. The characteristic activity of the working mathematician is to scribble something, "meaningless marks", which gives rise to thoughts, "languageless activity", which in turn leads to more scribbles, more thoughts, and so on.

We need not follow this road any further here, though it holds promise of taking the discussion about the nature of mathematical knowledge in a more fruitful direction than the classical triad tends to do. As educators we are more interested in the relation between the philosophies of mathematics and the practices of teaching mathematics. In the classroom formalism appears allied with symbol-pushing, two-column proofs, rote learning of skills, and other forms of making meaningless marks on paper. On the other hand intuitionism, where mathematics is seen as a languageless activity, supports an abstract, a-cultural approach to mathematics, and an unrealistic emphasis on understanding mathematics before doing it. What is not clear to me, though others seem more confident in their deductions, is the extent to which anyone's philosophical view has any causal effect on what happens in classrooms.

3. A SOCIOLOGY OF KNOWLEDGE

The uses to which the word "knowledge" is regularly put create certain associations. These often act like supplementary meanings of the word, meanings which are not included in its definition but which nevertheless give it a particular flavor, a particular emphasis. Some of these extra meanings can be discouraging to the seeker after mathematical knowledge as the following examples show:

"Knowledge is" ... anonymous, achieved by someone else ...
"Knowledge is" ... static; finished; polished ...
"Knowledge is" ... found in books, or in the minds of experts ...
"Knowledge is" ... measurable; one can have a lot or a little ...
"Knowledge is" ... authoritative; compelling; always right ...
"Knowledge is" ... pure; selfless; impersonal ...
"Knowledge is" ... like property, one must pay a price for it ...

No doubt other associations, other hidden meanings, will occur to readers. And while on the subject of the word itself, it is worth noting in passing that the English language does not allow one to speak the plural of knowledge, i.e., of "knowledges", though a francophone is able to speak of "connaissances". For many purposes the word "knowing" (which can be a noun as well as a participle in English) may be a more useful word, less encumbered with hints of difficulty, impersonality, and the like. The resemblance to the verb form keeps alive the personal connection; only people can know and hence produce knowing(s). We could not, of course, outlaw the word knowledge, even if we wanted to, but it does seem to be an important exercise to remind ourselves frequently that quite respectable words like knowledge may be carriers of meanings that we do not intend when we use them.

4. OUR EDUCATIONAL TARGETS

It would not diminish the importance given to mathematical knowledge if we placed it in a slightly wider context. As an arguing point, let me suggest that our specific aims in teaching mathematics should be to give every student the opportunity to:

- Function "like a mathematician".
- Develop a repertoire of general mathematical "know-hows".
- Appropriate some mathematical knowledge.

The primary goal is, I suggest, the first one. I do not confine it to those students who may become professional mathematicians or technologists. Its significance for me is that it says more exactly what seems to me to lie behind proposals for including mathematical problem solving, say, or mathematical investigations in the school curriculum — that in a modest way and at an elementary level everyone is capable of thinking and acting "like a mathematician does" and that schooling can help to educate the capacity. The second goal in the list is not more important than the third but I place it second because, as Edwina Michener (1978) has persuasively pointed out, it is often overlooked. Students in mathematics classes need a range of "know-hows" (how to simplify a calculation, how to check the solution of a problem, how to select a general case or a generic example, how to assemble a proof, how to construct a counterexample, for instance) and these know-hows are very rarely explicitly taught.

Whatever particular approaches to these educational goals are adopted, I believe it is important that they have the following two characteristics:

1. They give the students at least some limited experience in creating or modifying some mathematical knowledge. Without some of this kind of

experience, the students will not be able to take the step from accepting knowledge because someone tells it to them to accepting knowledge because they know it to be true.

2. They give the students more than one way to access mathematical knowledge so that the students can check and validate their own beliefs. Without this experience, the students will remain dependent on authority and unable to achieve autonomy in their mathematical work.

No one said mathematics teaching had to be easy!

5. KNOWLEDGE AND ASSESSMENT

It is not generally difficult to find out what one wants to know about what another person knows, given that other person's cooperation and enough direct access to be able to put questions. The problems of assessment derive almost entirely from a perceived need to assess everything and everyone involved in the learning of mathematics and to communicate results of the assessing to anyone who asks. A publicly accountable educational system in a complex mobile society, which is expected to take people with "who-knows-what" backgrounds and prepare them in vast numbers for "who-knows-what" interests, occupations, and further educational opportunities, requires some way of collecting and encapsulating information about individual students that can be made available to other people, usually strangers, both inside and outside the system. The scale of the enterprise raises genuinely difficult questions about assessment procedures, particularly how to make them accurate, equitable, informative, and credible. Unfortunately, the difficulties are usually not approached in the systemic way the size and complexity of the situation demands: A simple-minded model of assessment is often adopted, and procedures are introduced without anticipating their influence on the teaching-learning situation as a whole (though if people involved in education know *anything*, they should surely know the likelihood of that!).

It is unfortunate that elementary mathematical knowledge appears to lend itself so readily to assessment by simple-minded means. All the clichés about mathematical knowledge — "you either know it or you don't", "an answer is either right or wrong", "you just have to memorize standard facts and procedures", etc. — reinforce the belief that its definiteness, its lack of shades of meaning, make mathematical learning easy to assess. And when applied to the most elementary levels of mathematics, these clichés do seem to capture a truth about the quasi-mechanical fluency that is characteristic of the behavior of successful students. The lack of certainty and absolute finality that the mathematicians and philosophers talk about really does not seem very germane at the level of students who are learning to compute a product of fractions, to factor a trinomial, even to prove a simple geometrical theorem.

The real danger in the simple-minded view of mathematical assessment does not lie so much in its failure to do what it sets out to do as in the irresponsible way it overlooks its own consequences. I consider just one aspect here; others will certainly occur to any reader.

Mathematical Concepts and Skills

The cliché-driven view of mathematics learning is that there are certain basic concepts and skills that have to be acquired if one is ever going to be able to "do" mathematics or to "understand" it, so these must first be learned and "mastered". This way of picturing mathematics learning leads to *itemizing* the "basic content" of mathematics, identifying those concepts and skills that it is essential to know. This in turn implies an *atomization* of knowledge (since the concepts and skills must be delineated and differentiated from each other), and implicitly to a *finalization* of knowledge (since the items are to be mastered and will henceforth be in some sense fixed and finished). Now, it is possible that some mathematical skills can be learned once-for-all, needing only to be practiced and rehearsed after that, but concepts are only locally stable and are always potentially subject to change. But more important than this is the effect of the "concepts and skills" model on the learners, who usually have no access to a vision of mathematics other than the one their classroom provides. What are they to make of this activity which offers them a multiplicity of mismatched building blocks but never asks them to build anything? How can they know what one can do with mathematics? How can they invest this reductive activity with the sort of meaning a mathematician might see in it? For the students, it may easily seem as if mathematics is no more and no less than this collection of largely uncoordinated items of information and behavior.

The proven backwash effect of assessment procedures has to be taken into account. It is not sufficient for educational psychologists to devise procedures that are precisely tailored to measure learners' behaviors on items of de-contextualized mathematical knowledge. Assessment procedures need to be considered *in interaction with* both the institutionalized and the non-institutionalized features of mathematics education — the goals, the professed values, the curriculum, the teaching methods, and all the other aspects of the social and cultural context, large and small.

This is a considerable challenge. Here are some others:

- To assess all levels and forms of mathematical knowledge, including general mathematical "know-hows" as well as specific information and narrow "how-tos".

- To structure assessment so that it takes account of different levels of significance of mathematical knowledge, so that it can distinguish some achievements as more advanced than others.
- To provide different forms of information obtained from assessment. "Strangers" may need global information about a student's potential whereas the student needs feedback which is specific and locally helpful.

Assessment procedures would almost certainly fulfill their function just as well and be a lot less harmful in their effects if they utilized a variety of formats and were not afraid to examine similar behaviors more than once in different ways. *Redundancy* and *plurality* while not sufficient, are probably necessary features of a well-designed assessment system.

Taking account of this complexity may seem certain to compromise the "reliability" and "validity" of assessment procedures. But these properties have no intrinsic right to their places at the top of the heap of priorities: they are artifacts of a particular reductive "simple-minded" theory that essentially dismisses the interactive effects of assessment that we have mentioned as if they were not its concern. Everyone would probably benefit by settling for a somewhat reduced reliability and validity in order to gain, for example, a somewhat enhanced *equity* and *credibility*.

As for the assessment of mathematics per se, it seems a good time, now that educators have at last started talking seriously about the specific characteristics of mathematical knowledge, to try to shift "the assessor's epistemology" much closer to the epistemology of the mathematician or, better still, that of the informed, reflective, mathematics teacher.

REFERENCES

Castonguay, C.: 1972, *Meaning and existence in mathematics*, Springer-Verlag, New York.

Dauben, J.W.: 1979, *Georg Cantor: His mathematics and philosophy of the infinite*, Harvard University Press, Cambridge, MA.

Hadamard, J.: 1945, 1954, *An essay on the psychology of invention in the mathematical field*, Princeton University Press, Princeton, NJ (Republished as a Dover Book).

Kitcher, P.: 1983, *The nature of mathematical knowledge*, Oxford University Press, New York.

Michener, E.R.: 1978, 'Understanding understanding mathematics' *Cognitive Science*, **2**, 361–383.

Poincaré, H.: 1913, 1952, 1956, 'Mathematical creation. A paper delivered to the Psychological Society of Paris', in Poincaré, H., *Science and Method*, The Science Press, New York, republished as a Dover Book, also in Newman, J.R. (Ed.), *The World of Mathematics*: Vol. 4. Simon and Schuster, New York.

Rotman, B.: 1988, 'Towards a semiotics of mathematics', *Semiotica*, **72** (1/2), 1–25.

David Wheeler
White Rock,
British Columbia,
Canada

THOMAS A. ROMBERG

HOW ONE COMES TO KNOW: MODELS AND THEORIES OF THE LEARNING OF MATHEMATICS

1. INTRODUCTION

Charles Schultz (Courtesy United Features Syndicate, Inc.)

This cartoon humorously reflects the following facts: (1) Schools have been created by social groups to prepare their young for membership in society; (2) to this end, every society has determined what knowledge it expects its young to acquire (e.g., wanting Stripe to whistle); and (3) each society has made provisions for how students are to acquire that knowledge (i.e., by the curriculum advocated, teaching standards, etc.). Unfortunately, establishing such goals and procedures is no guarantee that learning will occur. Assessment of student performance (e.g., finding out whether Stripe can whistle) is the only way one can determine whether students have acquired the intended knowledge. Given this perspective, there are two fundamental epistemological questions that must be asked and answered.

- What knowledge do "we" want students to have?
- How does a student come to know?

M. Niss (ed.), Investigations into Assessment in Mathematics Education, 97–111.

It is critical that we have answers to these questions if reasonable assessment procedures are to be established. Also, both questions continually need to be asked and re-answered since social conditions change and our ideas about learning change. Without such reconsideration, societies risk teaching students ideas that are no longer useful for societal membership via methods that are no longer valid. This situation has been satirized in *The Saber-Tooth Curriculum* (1939) by J. Abner Peddiwell, who describes how students are taught to kill saber-tooth tigers after they have become extinct (comparable to the practice in some schools today of teaching students how to use a slide-rule or interpolate log tables).

In this chapter, my intent is to briefly summarize current thinking on the second of these two questions, How does a student come to know? and to draw implications from this summary for assessment. I do this with full awareness that throughout the world, the mathematical sciences community is involved in serious reconsideration of the first question. For example, the publication in England of *Mathematics Counts* (CITMS, 1982) and in the United States of *Curriculum and Evaluation Standards for School Mathematics* (NCTM, 1989) reflect similar visions of what the mathematical sciences community considers it important for all students to know. It is a vision of mathematical knowledge that is quite different from that of the past century. But, how its content is organized for instruction, the manner in which it is taught, and the way in which mathematical performance is assessed often fail to reflect those changes. This failure is due, in part, to not considering the second of these epistemological issues. To appreciate the difficulty in answering this question, I have chosen to briefly summarize six past theories of learning. Then I describe in more detail contemporary notions about learning. I conclude this chapter by reflecting on the implications of these latter notions about learning for the assessment of mathematical performance.

2. PAST THEORIES OF LEARNING

Notice the plural — theories of learning. There are many different theories, because how humans learn is extremely complex. First, there is no general agreement on a definition of learning. Some define it as "change in behavior", others as "acquisition of new knowledge", or "gaining a new understanding". In addition, there is no agreement on either the details of how learning takes place or what constitutes reasonable evidence that learning has taken place. The latter is necessary if theories are to be tested — i.e., validated, or falsified. Second, part of the reason for lack of agreement on a learning theory is that there are many different kinds of learning. For example, learning how an ellipse is defined and the form of an equation that can be used to represent it is different from learning how to solve simultaneous equations using matrices, and both are different from learning how to use trigonometric ideas to solve a surveying problem. Thus, depending upon the kind of learning a mathematics educator or a psychologist studies — concepts, skills, problem solving — a different theory may be developed. Third, learning theorists have made different philosophic assumptions about the nature of the learning process. These contrasting assumptions include the following: whether learning occurs by passive reflection or by active construction, whether limits of learning are biologically determined or environmentally determined, whether learning is a rational process or an irrational process.

Thus, given that there are different things to be learned and that theorists have made such differing assumptions about learning, it should not be surprising that they have proposed many different explanations of how learning occurs. In this paper, I have chosen to concentrate on six general theories that were developed during this century and currently influence the way the learning of mathematics is studied. By focusing on 20th century theories, one should not infer the paucity of theories prior to this century. Scholars as far back as Plato in the 3rd century BC have developed theories of learning, and some, like John Locke's atomistic model, have been precursors to current theories (Phillips & Soltis, 1991). Nevertheless, I have chosen to focus on the following six theories because each of the theories has been used to explain some aspect of mathematical learning during the recent past.

Behaviorism

This theory, which evolved at the end of the last century, focuses on the outcomes of learning (behaviors) rather than on how learning occurs. It assumes learning occurs by passively, but rationally, reflecting on stimuli from the environment. In fact, it conceives of all psychological phenomena as behavior stimulated by the environment, not as mental processes. The basic assumption of this theory is that a child is like a machine that must respond to the stimuli of an outside force and increasingly conform in its responses to the demands of that force. The theory goes on to address the mechanisms for transmitting environmental content to the child (respondent and operant conditioning) and the formation of associations, which constitute the growth and memory of the child (Langer, 1969). This theory has been used by scholars to study how desired responses to stimuli (outcomes) become fixed by practice and praise (reinforcement). Learning is viewed as change in behavior (or performance) and change scores (pre- versus post-test differences) on measures of performance, based on an increased number of correct answers to specific questions, are often used as evidence for learning. This theory, in its many forms, has strongly influenced all education in the United States and mathematics education, in particular. From the publication of Edward L. Thorndike's *The Psychology of Arithmetic* (1922) through Robert Gagné's *The Conditions of*

Learning (1965), mathematics educators have utilized principles drawn from behaviorism. Drill-and-practice routines (e.g., to memorize multiplication facts, or factor trinomials, or solve linear equations) follow the tenets of this theory.

Differential (or Mediational) Behaviorism

In the 1930s, a variant of behaviorism was developed in response to a major flaw in classical behaviorism. It had been assumed that different individuals would exhibit the same behaviors if they were given the same stimuli and reinforcements. However, data failed to support this assertion. Some individuals learned faster and retained more than others, as well as exhibiting other variations in learning. To explain these variations, many hypotheses were posed and examined (Anastasi, 1958). Explanations gradually focused on genetically determined biological or neurological differences in how information was absorbed, arguing that individuals differ in learning because of predetermined aptitudes, abilities, capacities, or styles which mediate how information is processed. One major line of research based on this theory has been "aptitude-by-treatment interaction (ATI) studies" (Cronbach & Snow, 1977). In such studies, students with different aptitudes receive different treatments. Schools today form ability groups and tracks, for instance, based on these ideas. Again, learning involves a change in performance, and change scores on standard tests constitute the usual evidence of learning for this theory.

Gestalt Constructivism

While behaviorism, in its many forms, was an adequate model for many forms of learning (particularly the learning of low-level concepts and skills), it was totally inadequate in explaining how one discovered a relationship, proved a theorem, and solved complex problems. Repeated practice and reinforcement cannot make someone a creative mathematician; these strategies do not foster the invention of new ideas. Learning does not consist simply of a change in performance. Therefore, the increases in the number of correct answers to simple tasks fail to capture changes in strategies or ways of thinking about a problem. Thus, in the 1930s, *Gestalt* theory developed. The German word "Gestalt" means organization or configuration (Hilgard & Bower, 1975). This theory posits that learning involves active construction, rather than passive absorption from the environment, and it implies that students experience the world in meaningful patterns and then construct meanings from those patterns. For this theory, evidence of learning is registered in terms of changes in how persons said they thought about problems (coded from think-aloud interviews). The best examples of this approach can be found in the work of Wertheimer (1959), who was interested in the problem of "productive thinking" in mathematics. Most early work on mathematical problem solving (e.g., Polya, 1945; Hadamard, 1945) is rooted in this theory. Also, in contrast to the behaviorists, the place of meaning was of particular interest to these psychologists.

Developmental Constructivism

As differential behaviorism grew out of a need to explain why persons differ in responses, developmental constructivism grew out of a need to explain why the same persons at different ages constructed different meanings from the same environment. The explanation for this phenomenon, credited in large part to Jean Piaget, also was based on biological traits. Piaget argued that human learning differs and changes in various evolutionary stages, which he called preoperational, concrete operational, and formal operational (Flavell, 1963). The influence of this theory has been particularly strong on the study of young children's learning; it is now well-documented that they see and operate in the world differently from adults. Learning, from this perspective, involves the restructuring of knowledge. Furthermore, change or restructuring is dependent upon the notion of maintaining balance or equilibrium and the processes of assimilation and accommodation needed to maintain that balance. Evidence of learning for Piaget was based on this *méthode clinique* — a method which involved interviews of children as they worked on challenging sets of problems.

Dialectic Constructivism

In contrast to Piaget's explanation of differential construction over time, Soviet psychologists (notably, Luria, 1979; Vygotsky, 1962; and Leont'ev, 1981) posed an alternate theory according to which imbalance, not equilibrium, is considered normal (Riegel, 1979). Change — in a student's language development, in particular — is dependent on dialectic synthesis, on one's environment, and on the actions or activities in which the subject is involved. Again, evidence is typically based on data from interviews of individual subjects. Interest in this theory for mathematics teaching is in part due to the publication of the 14-volume *Soviet Studies in the Psychology of Learning and Teaching Mathematics* (Kilpatrick & Wirszup, 1969–1975), and the publication in English of Krutetskii's *The Psychology of Mathematical Abilities in School Children* (1976), and Davydov's *Types of Generalization in Instruction* (1990).

Psychoanalytic Behaviorism

This theory, which assumes that learning is based on the irrational desires, fears, and other emotions of humans, is rooted in the work of Sigmund

Freud at the turn of the century. According to this theory, conditioned reflexes do not adequately explain many of the behaviors of children, like imitation, or wanting specific things. It maintains that covert responses attributable to the child mediate production of other overt responses (Langer, 1969). It assumes that basic drives (hunger, sexuality, self-concept) are at the basis of learning. Humans construct attitudes and perspectives based on these drives. Erikson (1968) proposed stages of development from this perspective. Evidence supporting this theory comes from clinical psychoanalysis (which involves a special form of clinical interview). This theory has become a prominent factor in mathematics education in the study of mathematics phobias and similar phenomena.

3. CONTEMPORARY NOTIONS ABOUT LEARNING

Although each of the above six general theories has contributed significantly to our understanding of the learning of mathematics, most current psychological research is based on very recent notions about how humans process information. Behaviorists ignored processing; differentialists argued for global aptitudes; and others gave other general biological or environmental explanations for how learning takes place.

Contemporary psychological models are now beginning to emerge — the outgrowth of a revolution in psychology which has been going on during the past quarter-century.[1] Although there is not complete agreement on details among the different psychological schools, I have decided to label the models as "social constructivist" because they see learners as active constructors of knowledge in a social environment. The social constructivist models evoke eight common features of how the mind works. These generally agreed-upon features draw upon research that encompasses ideas derived from earlier models: social negotiation from Vygotsky and his followers, memory structures from artificial intelligence, schema structures from the cognitive sciences, levels of reasoning from post-Piagetian developmentalists, and so forth.[2]

The eight features of this social constructivist view of learning are:

1. There are key components of mental function that are assumed to be related as follows (see Figure 1).

Processing begins with an experience. Information from the experience is filtered by an executive control mechanism that involves deciding what to do with the information derivative from the experience and the beliefs and attitudes held with respect to that information. Next, mediated information is organized and stored in memory. This critical aspect of cognitive theory distinguishes between three types of memory: working memory, short-term memory, and long-term memory. Working memory

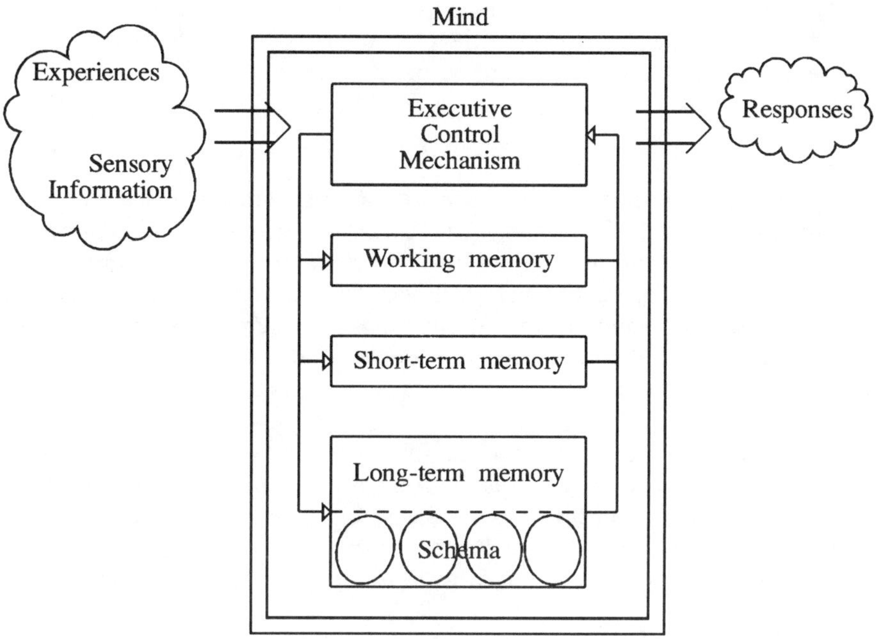

Figure 1

involves that limited amount of information that one can actually process at a particular point in time. The units of information may be individual pieces or may be related pieces (chunked information). Information is stored in short-term memory if it is being used in a relatively short period of time. Then it is either dumped (forgotten) or is stored in long-term memory.

2. While humans are capable of remembering a great deal, they have an extremely limited capacity to think about several different things at any one time. As a consequence of this limited capacity, information stored in long-term memory must be well organized. The mind naturally organizes repeated similar experiences in long-term memory into what psychologists call *schema*. These are complex networks of concepts, rules, and strategies, not isolated facts or procedures. For example, one typically organizes the experience of eating in restaurants into a "restaurant schema". The general features of past experiences are organized into a network of related facts (e.g., waiters, tablecloths, menus) and relationships (e.g., sequence of ordering, serving, paying). Having information stored in this way helps an

individual cope with new experiences. For example, when one is in a new city and sees a sign for a restaurant, one "knows" what to expect based on past experiences. Such schema develop over long periods of time and by continual exposure to related contextual events. For the learning of mathematics, Gerard Vergnaud's (1982) notions about *conceptual fields* reflect these ideas.

3. New experiences either use one's existing schema — a process termed *assimilation* — or force a change in a particular schema, termed *accommodation*. For example, if one had never been to a Chinese restaurant, such an experience (e.g., shared dishes, chopsticks, and so forth) would force one to add new features about restaurants to one's schema. Also, note that new experiences are often social and depend in large part on language experiences with others.

4. Naturally occurring schema are idiosyncratic to the individual, who is usually unaware of the organization. For example, each individual's restaurant schema differs from others' because of different restaurant experiences, but one rarely thinks about the consequences of persons having different schema related to the same general phenomena.

5. While most learning occurs through the natural assimilation and accommodation of experiences, it can also occur via pre-organized, structured experiences in which groups have an opportunity to discuss what is happening. Much of what is presented in school should have this character. Instructional activities should serve as the means of connecting one's informal, natural experiences with the formal aspects of any discipline via communication.

6. Coherent instructional activities that are designed to promote conceptual reorganization in an individual's schema should have a three-part sequence. First, there should be an *exposing event* which encourages students to use and explore their own conceptions in an effort to understand the event. This is followed by a *discrepant event*, which serves as an anomaly and produces cognitive conflict. It is hoped that this will lead the students to a state of dissatisfaction with their current conceptions. A period of *resolution* follows in which the alternative conceptions are made plausible and intelligible to students, and in which the students are encouraged to make the desired conceptual shift (Nussbaum & Novick, 1982). This sequence should be seen as a negotiation process between students and instructors. The objective is for students to develop well-organized schema that reflect current notions about the discipline. It is also argued that the encoding, comprehension, and retrieval of new information are aided when material is presented in a form that has structure and when the student is cognizant of that structure. In particular, these processes are facilitated when the

information can be related to existing schema of the learner. Thus, when information is presented in a familiar contextual setting, the transitions, concepts, and procedures are likely to be remembered. Some psychologists call this a generic story shell for the schema.

Educators must be aware that the negotiation process is critical because many (if not most) experiences students have had are related to the world outside of school. Thus, school experiences should build on the notions students bring to the classroom, provide a rich set of new experiences so that all students have some experiences in common, and provide a social environment (a discourse community) in which common meanings, interpretations, and justifications for conjectures about mathematical domains can be negotiated. Thus, knowledge should be acquired by students via active, social construction.

7. Schema are never fixed. They continually change over time as the individual grows and has additional experiences. In fact, there is general agreement among developmental psychologists that the evolution of thinking goes through several cyclic stages. A number of these psychologists have proposed similar theories which trace the stages of student development from early psychomotor reactions to logical thinking (e.g., Biggs & Collis, 1982; Case, 1985; Fischer, 1980). The elements that these theories have in common enable us to use their insights as a sound basis for identifying key elements in the achievement of understanding and competence during the school years.

(a) Two phenomena are involved in determining the level of an individual's response to an experience — *mode* of functioning, which is determined by the level of abstraction of the elements utilized, and the complexity of the *structure* of the response within that mode.

(b) There are at least four modes of functioning. The elements involved in the *sensori-motor mode* are the objects in the immediate physical environment and the operations on them are concerned with their management and coordination. In the *iconic mode*, the elements become signifiers (words or images) which stand for objects and events; the operations involve the manipulation of these signifiers to establish oral communication, to make links between affect and image, and to make perceptually based qualitative judgements. In the *concrete-symbolic mode*, the elements are not just signifiers, but concepts that are manipulated using a logic of classes and equivalences, both elements and manipulations being directly tied to the empirical world. This mode enables the concrete world to be interpreted through symbolic systems such as written language and the signs and symbols of mathematics. Finally, in the *formal mode* the elements become abstract concepts and propositions and the operations on them are concerned with determining the actual and deduced relationships between them; neither the elements nor the operations need a real world referent.

(c) The modes progress from concrete actions to abstract concepts and principles, and the emergence of one mode does not replace its predecessor. The modes, in fact, are cumulative — the later-developing modes existing along with the earlier modes. The implications of this last statement are that as an individual matures physiologically, the modes developed earlier continue to develop as a consequence of both physical maturation and experiences and, as the modal repertoire available increases, multi-modal functioning becomes the norm.

For novices, it is important to note that they do not, as yet, think like experts. They are more likely to respond in an iconic or concrete-symbolic mode to new information. Only as their repertoire of experiences broadens do they begin to build formal relationships between disparate experiences; but even as yet, they do not think in terms of composite, abstract concepts.

8. Students who have formed initial and not well-organized schema actively search for experiences that will provide them with structure. However, people resist changing well-developed schema because such schema have great assimilative power. What is supposed to be a new view may be assimilated by the old. It is difficult for teachers to distinguish between assimilation and accommodation. People whose important beliefs are threatened will attempt to defend their positions, dismiss objections, ignore counter-examples, keep logically incompatible schema segregated, and so forth.

Thus, we see that these social constructivist notions about learning are quite different from the strictly behaviorist notions of learning that have most strongly influenced education in the United States. Elsewhere, this contrast has been summarized by Romberg, Zarinnia, & Collis (1990) as follows:

"All learning involves thinking, but in the past, most instruction focused on learning to name concepts and follow specific procedures. Now the emphasis for all students must shift to communication and reasoning skills. Although these skills resist precise definitions, they are now popularly called "higher order" thinking skills. Lauren Resnick (1987) listed some of their features, many of which are in stark contrast to current mathematics criteria (shown in parentheses).

- Higher order thinking is *nonalgorithmic*. That is, the path of action is not fully specified in advance (still largely algorithmic).

- Higher order thinking tends to be *complex*. The total path is not "visible" (mentally speaking), from any single vantage point (standard examples with visible paths).

- Higher order thinking often yields *multiple solutions*, each with costs and benefits, rather than unique solutions (single, unique solutions).

- Higher order thinking involves *nuanced judgement* and interpretation (neither judgement nor interpretation expected).

- Higher order thinking involves the application of *multiple criteria*, which sometimes conflict with one another (simplified to single criteria that are well defined in content).

- Higher order thinking often involves *uncertainty*. Not everything that bears on the task at hand is known (certain — all information required is given).

- Higher order thinking involves *self-regulation* of the thinking process. We do not recognize higher order thinking in an individual when someone else calls the plays at every step (external regulation).

- Higher order thinking involves *imposing meaning*, finding structure in apparent disorder (meaning is given or assumed).

- Higher order thinking is *effortful*. There is considerable mental work involved in the kinds of elaborations and judgements required (work which usually involves standard exercises is simplified so that little effort is needed).

Given these contrasts, then, thinking skills must be the focus of instruction in mathematics in the near future, and assessment procedures need to be developed that portray not only the number of correct answers students can produce, but the thinking that produced those answers." (pp. 22–23).

4. ASSESSING MATHEMATICAL PERFORMANCE

The concept of "well-organized schema" is now used to explain how people impose order on experiential information. Thus, the assessment question is "How does a child order new information and how does such ordering change over time?" Obviously, assimilation, accommodation, and mode of functioning in response to new information are important in the enterprise of schooling. However, as new information is encountered, it is important to understand how a child actually processes that information. Learning by active participation implies that instruction should be based on the kinds of activities from which a knowledge of skills can be developed. However, simply developing a collection of interesting activities is not sufficient. The knowledge gained must lead somewhere. Again, it would be important to know whether new information has led to a restructuring of knowledge. The knowledge that is constructed by any individual depends to some extent on what is brought to the learning situation, where the current activity fits in a sequence leading toward a goal, and how it relates to society's definition of knowledge in that field. Thus, the suitability and effectiveness of selected learning activities is an empirical problem. Its evaluation depends on the student's prior knowledge, the student's expectations, and society's expectations. One can only determine this from information about student performance.

These notions are important since, without a schema into which new information can be assimilated, experience is incomprehensible, and, therefore, little can be learned from it. But the schema by which a student assimilates a lesson may not be that assumed by teachers. This fact can easily escape detection since the student will often be able to repeat segments of the text and lecture even though s/he understands them in terms of an incorrect, incomplete, or inconsistent framework. Indeed, students may develop specialized frameworks for maintaining the particular identity of lesson material in order to cope with the demand for veridical reproduction. Schema use must be a dynamic, constructive process, for it is not the case that people store a schema to fit every conceivable situation. In this view, the acquisition of knowledge implies schema change, not just the mere aggregation of information.

The problem now to be faced is how to develop assessment procedures that will inform teachers about schema change. To approach a solution to this problem, Antoine Bodin elsewhere in the present book argues that the purpose of assessment is to organize (or to scrutinize) situations in such a way that it enables us to gather information which, after processing, can reveal something that is consistent about the problem by characterizing personal knowledge in terms of networks for specific mathematical domains (schema for a domain).

To illustrate the problem, I have chosen to contrast three modes of assessing student performance and the assumptions about what it means to know mathematics (i.e., what an examination of information will reveal). Each mode of assessment begins with a student being expected to respond to one or more tasks and is designed to examine the student's responses.

The standardized *norm-referenced* test based on behaviorism assumes that one knows isolated pieces of knowledge. Thus, such a test asks students to respond to a variety of questions about specific parts of mathematics, some of which the student knows and some not. Responses are processed by summing the number of correct responses to indicate how many of the parts of mathematical knowledge a student possesses and the totals for an individual student compared to those of other students.

Criterion-referenced assessment, a second common form of organizing situations for judging student performance, is also based on behaviorism. However, it asks students to respond to specific sets of questions about specific parts of mathematical knowledge so that the level of mastery of each part can be judged. This yields for each student a profile of those areas of knowledge that have been mastered. The problem with both approaches is that neither yields information about the interrelationships among the parts of knowledge (the structure of the schema) held by a student. Also, neither approach reveals anything about schema change.

Authentic performance assessment (Lajoie, 1992), based on constructivist notions, begins with complex tasks which students are expected to work on for some period of time. Also, their responses are not just answers; instead, they are arguments which describe conjectures, strategies, and justifications. This yields for each student a picture of the structure of schema. It is this type of assessment many around the world are now working toward.

To assess personal knowledge from a constructivist perspective and how it changes will involve the development of authentic assessment strategies. This includes specifying what well-formed schemas about the basic mathematical domains would be like, developing tasks for students to respond to that both measures what parts of each domain a student knows and how the student has connected them; it also involves repeated assessments to determine change in both, and so forth. In addition, the development of such strategies will involve new ways of determining validity and reliability, and new ways of reporting results.

Authentic assessments can be developed by determining the degree to which each student has grown in his or her ability to solve nonroutine problems, to communicate, to reason, and to see the applicability of mathematical ideas to a variety of related problem situations.

5. CONCLUSION

The assessment challenge we face is to give up old assessment methods to determine what students know which are based on behavioral theories of learning and develop authentic assessment procedures that reflect current epistemological beliefs both about what it means to know mathematics and how students come to know.

NOTE

1. This follows Paul Ernest's (1991) discussion of the parallelism between the philosophic notions about social constructivism and those from psychology.

2. I have drawn heavily on the following works in building this summary in addition to those sited directly (Silver, 1987; Schoenfeld, 1985, 1987; and Hiebert & Carpenter, 1992).

REFERENCES

Anastasi, A.: 1958, *Differential psychology: Individual and group differences in behavior* (3rd ed.), Macmillan, New York.

Biggs, J.B. & Collis, K.F.: 1982, *Evaluating the quality of learning: The SOLO taxonomy*, Academic Press, New York.

Brownell, W.A.: 1972a, 'The place of meaning in mathematics instruction: Selected research papers', in Weaver, J.F. & Kilpatrick, J. (Eds.), *Studies in mathematics*, **22**, School Mathematics Study Group, Stanford, CA.

Brownell, W.A.: 1972b, 'The place of meaning in mathematics instruction: Selected theoretical papers', in Weaver, J.F. & Kilpatrick, J. (Eds.), *Studies in mathematics*, **21**, School Mathematics Study Group, Stanford, CA.

Calfee, R.: 1981, 'Cognitive psychology and educational practice', in Berliner, D.C. (Ed.), *Review of research in education* (3–74), American Educational Research Association, Washington, DC.

Case, R.: 1985, *Cognitive development*, Academic Press, New York.

Committee of Inquiry into the Teaching of Mathematics in Schools: 1982, *Mathematics counts* (the Cockcroft Report), Her Majesty's Stationery Office, London.

Cronbach, L.J. & Snow, R.E.: 1977 *Aptitudes and instructional methods: A handbook for research on interactions*, Irvington, New York.

Davydov, V.V.: 1990, *Soviet studies in mathematics education: Vol. 2. Types of generalization in instruction*, National Council of Teachers of Mathematics, Reston, VA.

Erikson, E.H.: 1968, *Identity, youth, and crisis*, Norton, New York.

Ernest, P.: 1991, *The philosophy of mathematics education*, Falmer Press, London.

Fischer, K.: 1980, 'A theory of cognitive development: The control and construction of hierarchies of skills', *Psychological Review*, **57**, 477–531.

Flavell, J.H.: 1963, *The developmental psychology of Jean Piaget*, Van Nostrand, New York.

Gagné, R.: 1965, *The conditions of learning*, Holt, Rinehart, & Winston, New York.

Gardner, H.: 1985, *The mind's new science: A history of the cognitive revolution*, Basic Books, New York.

Greeno, J.G.: 1987, 'Mathematical cognition: Accomplishments and challenges in research', in Romberg, T.A. & Stewart, D.M. (Eds.), *The monitoring of school mathematics: Background papers: Vol. 2. Implications from psychology; outcomes of instruction* (3–26), Wisconsin Center for Education Research, Madison.

Hadamard, J.: 1945, 1954, *The psychology of invention in the mathematical field*, Dover, New York.

Hiebert, J. & Carpenter, T.P.: 1992, 'Learning and teaching with understanding', in Grouws, D.A. (Ed.), *Handbook of research on mathematics teaching and learning*, Macmillan, New York.

Hilgard, E.R. & Bower, G.H.: 1975, *Theories of learning* (4th ed.), Prentice-Hall, Englewood Cliffs, NJ.

Kilpatrick, J. & Wirszup, I.: 1969–75, *Soviet studies in the psychology of learning and teaching of mathematics* (Vols. 1–14), University of Chicago Survey of East European Mathematics Literature, University of Chicago Press, Chicago.

Krutetskii, V.A.: 1976, *The psychology of mathematical abilities in school children*, translated by Joan Teller, edited by Kilpatrick, J. & Wirszup, I., University of Chicago Press, Chicago.

Lajoie, S.: 1992, 'A framework for authentic assessment in mathematics', *NCRMSE Research Review*, **1**(1), 6–12.

Langer, J.: 1969, *Theories of development*, Holt, Rinehart, & Winston, New York.

Leont'ev, A.N.: 1981, 'The problem of activity in psychology', in Wertsch, J.V. (Ed. and Trans.), *The concept of activity in Soviet psychology*, Sharpe, Armonk, NY.

Luria, A.R., Cole, M. & Cole, S. (Eds.): 1979, *The making of mind: A personal account of Soviet psychology*, Harvard University Press, Cambridge, MA.

National Council of Teachers of Mathematics: 1989, *Curriculum and evaluation standards for school mathematics*, Reston, VA.

Nussbaum, J. & Novick, S.: 1982, 'Alternative frameworks, conceptual conflict, and accommodation: Toward a principled teaching strategy', *Instructional Science*, **11**, 183–200.

Peddiwell, J.A.: 1939, *The saber-tooth curriculum: A satire on American education*, McGraw-Hill Book Company, Inc, New York and London.

Philips, D.C., & Soltis, J.F.: 1991, *Perspectives on learning* (2nd ed.), Teachers College Press, New York.

Polya, G.: 1945, *How to solve it: A new aspect of mathematical method* (1957, 2nd ed.), Doubleday, New York (1971, reprint), Princeton University Press, Princeton, NJ.

Putnam, R., lampert, M. & Peterson, P.L.: 1990, 'Alternative perspectives on knowing mathematics in elementary schools', in Cazden, C. (Ed.), *Review of research in education 16*, American Educational Research Association, Washington, DC.

Resnick, L.B.: 1987, *Education and learning to think*, National Academy Press, Washington, DC.

Riegel, K.F.: 1979, *Foundations of dialectical psychology*, Academic, New York.

Romberg, T.A. (Ed.): 1991, *A Blueprint for Maths in Context: A connected Curriculum for Grades 5–8*, NCRMSE, Madison, WI.

Romberg, T.A., Zarinnia, E.A. & Collis, K.F.: 1990, 'A new world view of assessment in mathematics', in Kulm, G. (Ed.), *Assessing higher order thinking in mathematics*, 21–38, American Association for the Advancement of Science, Washington, DC.

Schoenfeld, A.H.: 1985, *Mathematical problem solving*, Academic Press, Inc., San Diego, CA.

Schoenfeld, A.H.: 1987, *Cognitive science and mathematics education*, Lawrence Erlbaum Associates, Hillsdale, NJ.

Silver, E.A.: 1987, 'Foundations of cognitive theory and research for mathematics problem solving', *Cognitive Science and Mathematics Education*, Lawrence Erlbaum Associates, Hillsdale, NJ.

Thorndike, E.L.: 1922, *The psychology of arithmetic*, Macmillan, New York.

Vergnaud, G.: 1982, 'Cognitive and developmental psychology and research in mathematics education: Some theoretical and methodological issues', *For the Learning of Mathematics*, 3(2), 31–41.

Vygotsky, L.S.: 1962, *Thought and language*, In E. Hanfmann & G. Vakar (Eds. and Trans.). Cambridge, MA: MIT Press.

Wertheimer, M.: 1959, *Productive thinking*, Harper & Row, New York.

Thomas A. Romberg
National Center for Research in Mathematical Sciences Education
University of Wisconsin-Madison
USA

ANTOINE BODIN

WHAT DOES TO ASSESS MEAN? THE CASE OF ASSESSING MATHEMATICAL KNOWLEDGE

1. INTRODUCTION

This chapter takes stock of lacks and weaknesses revealed in current assessment procedures. It also takes stock of the prospects opened by recent research. There are no questions of marks, examinations, or norm-referenced tests in this chapter. For these kind of situations, docimological research and criticism has for a long time provided us information that has more often than not been ignored. This chapter stresses the obstacles that are encountered when you want to gather reliable information (consistent, valid, fidel ...) on students' competencies and the fact that what is worthy of interest is the meaning of the observations rather than their numerical value. Criticism of assessment procedures or devices which belong to an objective test still constitutes a significant improvement over more traditional procedures. *EVAPM* studies are used as a base of operations for a part of the article. Some of the arguments are borrowed from a talk given at ICMI 6 (Budapest) with the title "Evaluation in Mathematics: the Quality of Knowledge".

When I was asked to give a paper that would answer the question "What does assess mean?" I did not realize on the spot how pretentious it would be for me to expect to give this question a comprehensive answer. Since the dawn of the this century, but with much emphasis during the last years, plenty of answers have been given in a lot of excellent works. Why should I or why would I want to add one more answer? In fact, I must confess that I have plotted to do it by roaming around the question more than dealing with it, as if I had not really understood the question or as if it were not addressed to me. Indeed, the question is very general and one may have to assess a person, a feeling, a situation, an organization, or a system to make a judgement about them. In that way, what is assessed may be the general functioning of the entity observed or a particular aspect concerning it or him. The scope is large and it would certainly be difficult and somewhat useless to seek a universal definition of the verb, *assess*.

Actually I will restrict my article to the educational context, and in this context I will concentrate exclusively on assessment about knowledge. Moreover, I will only consider cases where mathematical instruction is the

M. Niss (ed.), Investigations into Assessment in Mathematics Education, 113–141.

issue. Despite these restrictions, I regret that I do not have enough time and enough comprehensive knowledge of the topic to situate my personal answer according to the numerous answers already given (see references). So, about the given question, I first would ask: "Who is asked?" When assessing in an educational context, as in another didactical situation, we are concerned with at least three kinds of people and things: students, teachers (or assessors), and knowledge.

It is possible to answer the question as a student, as a teacher, or even as the knowledge itself, from a purely epistemological point of view. It is also thinkable to enter the question from an external, more or less specialized point of view. Indeed, answers given by a sociologist or by a psychoanalyst would not be the same. But even the answer given by a researcher interested in the learning of mathematics would likely not be the same as that given by a teacher or a student. Some didacticians who are mainly concerned with the study of the teaching/learning process would say: "Assessing is something imposed from outside of the didactical relationship which comes to disturb the phenomenon we want to observe". Figure 1 is shown to remind us of the stressing presence of assessment in the educational context. The whole assessment network appears as a system superposed over the teaching/learning system at its different levels (individual, class, school ...). Finally, I will present a personal and didactical point of view that can be contrasted with the other ones.

The Context in Which Our Observations Have Taken Place

We should make it clear that, though our observations were numerous and varied, they have taken place within a particular system: the French secondary teaching system where there are no examinations to spread the students through the different streams. From the beginning of Kindergarten until the end of schooling in high school, the rule is that of continuous estimation by the teacher of the class, this teacher not having to account to anyone, being free to choose as he wishes the rhythm, the contents, and the criteria of his assessment. The uses of marks to translate assessments is practically systematic, not only for the assessment of what the pupils produce, but also for the assessment of their global level. Determining who can move form one class to the next is done by calculating means; the average of the marks given to the productions (school work) is assumed to be equal to the achievement of their producer. On the contrary, what is the rule in some other countries, standardized tests and in a more general way tests externally produced, are quite unknown as with multiple-choice item formats.

Although in high schools the scientific sections dominated by mathematics and physics have the higher prestige (for parents as well as for teachers), over the last few years it has become difficult to fill these classes, despite their correspondence with the alleged needs of the society. Either

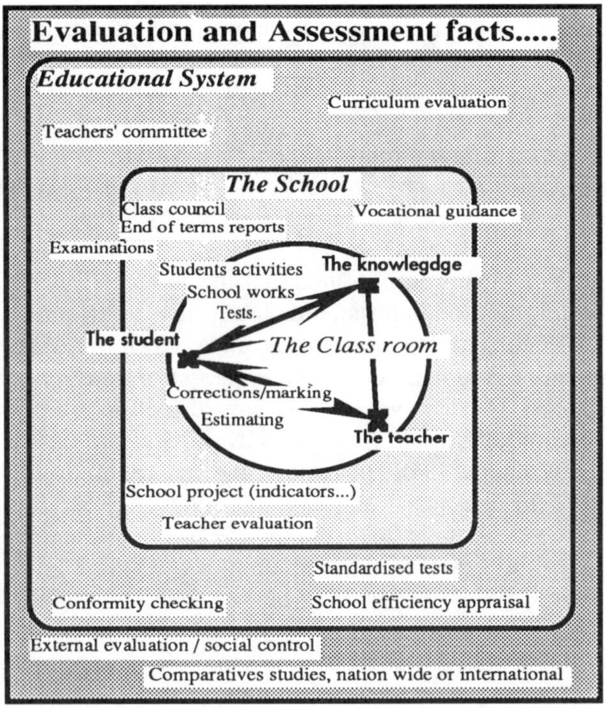

Figure 1

the students are not judged to be sufficiently good to enter these sections or they decide, of their own accord, not to apply to such classes.

2. ACCEDING TO KNOWLEDGE: THE STAKES IN ASSESSMENT

We are particularly interested in knowledge, or more specifically, in what it is possible to say about the knowledge of a person or of a group with respect to a particular area of knowledge. One of the inspirations for our research is our interest in the diffusion of knowledge. It is well known that this diffusion is limited by the idea that teachers have of the possibilities of their students, as well as by the idea that students have of their own capacities and their tastes. Without wishing to restrict the freedom of choice of people, one cannot ignore the fact that students' perceptions, to a large extent, result from teachers' judgements. Whatever the case may be, no one doubts that the procedures and the processes of assessment are largely responsible for the phenomena of exclusion.

3. TO ASSES: A POSSIBLE DEFINITION

I will risk a personal definition in this chapter and attempt to illustrate this definition and provide some clues on how to make it operational:

To asses means to organize (or to look at) a situation in such a way that it enables us to gather information which, after processing, can reveal something that is reliable about personal knowledge (or about the collective knowledge of a group).

A consequence of this definition is that the main concern of assessment should be saying something reliable about the individual or collective knowledge. This consequence would ask for a definition of the knowledge itself but, indeed, this is hazardous challenge that I am not ready to accept. For the moment, I will have to limit my definition to saying that any assessment is underlaid by a particular perception of the nature of knowledge. Any deeper thought concerning the reliability of a means of assessing leads to the question of what knowledge is. It is hardly an exaggeration to say, that, in the present state of the art, assessing demands the making of faithful reports about an unknown entity. Finally, I think that there is a dialectical link between our perception of what knowledge is and about possibilities for assessing it.

Whatever definition is used, someone's knowledge, fortunately for itself and unfortunately for its assessment, is not set once and for all. Knowledge essentially is changeable, subject to rearrangements, and more often than not, appears in disguised form. You seldom have the possibility of obtaining it directly, but may only find it revealed across casual types of situations or in situations organized on purpose.

The largest arrow in Figure 2 is aimed to show that what is most important is the progress of the teaching/learning process. This is true even when assessments are made for summative or certificative purposes (for instance examinations). If it were not, to use a metaphor, you would risk taking into account only the position of the observed phenomenon, paying no attention to its speed and still less to its acceleration.

The boldest arrow at the bottom of Figure 2 reminds us that decisions about educational context are fed not only from the knowledge-centered analysis. Numerous other deciding factors can interfere, factors that are linked with school policy, teachers' perceptions, etc., making it necessary to split off those factors in order to clarify the assessment process, rather than keeping them mixed and indistinguishable. But in this article, as I have already warned, I will only consider knowledge.

As a habit, we are using the word *competence* to point out a particular piece of knowledge that has been identified as well as possible. The word competence contains an idea of *potential*; although competencies are never observed directly, they may or they may not appear according to the

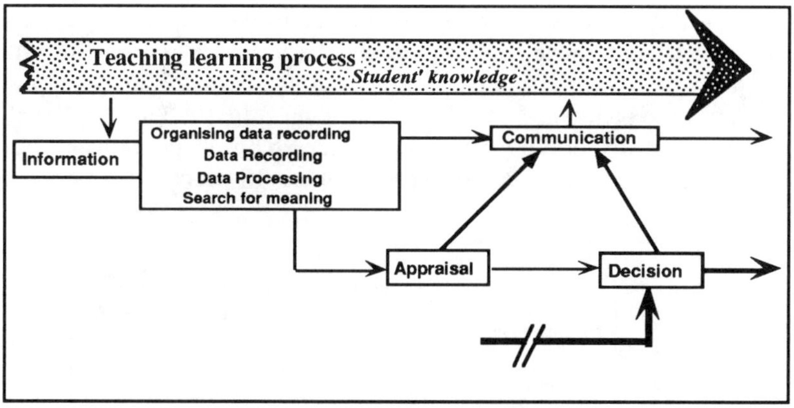

Figure 2

situations. Besides, what we said about knowledge may be transferred immediately to competencies. Surely the main difficulties in assessment concern the identification and limitation of competencies and they are linked with concerns about validity. Part of our observations come from the research conducted since 1981 at the IREM (Institut de Recherche sur l'Enseignement de Mathématique) of Besançon. Others come form the program called EVAPM.

4. THE EVAPM STUDIES

Since 1987 we have, with the French Mathematics Teachers Association, undertaken a large-scale assessment of the mathematical curriculum of the lower secondary and first year of upper secondary levels. These studies involved thousands of teachers and classes.

The aims of this program are at least fourfold:

- to follow the implementation of a new syllabus at lower secondary levels (6th Grade to 12th Grade).
- to help the general monitoring of mathematics teaching (curriculum development, etc. ...).
- to provide teachers with reliable indicators of students' competencies.
- to set up a permanent mathematical curriculum observatory.

In this chapter, questions coming from those studies are labelled EVAPM followed by the level observed: 6 means "Sixième" (6th Grade level), 5 means "Cinquième" (7th Grade level), etc. The year when the assessment took place is also indicated: 89 means 1989, etc. The percentages given on the pictures are those for correct answers, and the percentages given in italics are those for "non-answers" (N.A.). When useful, an item label is also given (e.g., Item D12). For instance, Figure 3 shows a question settled successively at level 5 in 1988, at level 3 in 1990, and at level 2 in 1991. The rates of good answers were successively 12 percent, 40 percent and 58 percent. The rates of non-answers were 16 percent, 13 percent and 02 percent.

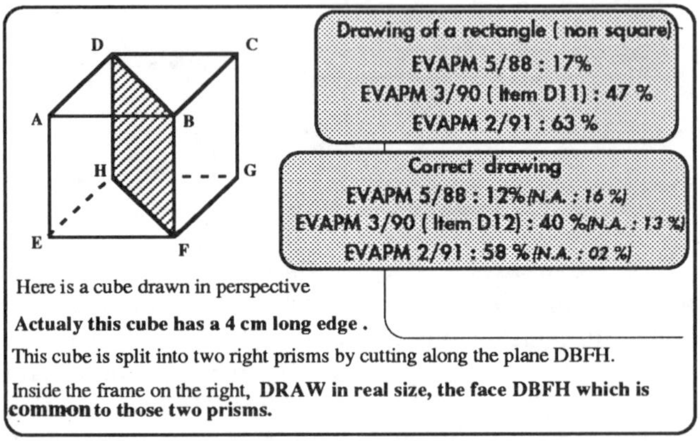

Figure 3

Actually, all the students tested at level 3 in 1990 did not enter level 2. For this particular question (the cube split in two prisms), 42 percent of the students who enter level 2 had given a good answer (instead of 40 percent for the whole population). This means that this question does not discriminate. For some questions, the gap is higher and may reach up to 40 percent.

All the students do not take the same tests, but we managed to compare (more exactly to cross) the results because questions were laid in the same test or in different tests. For instance, for item D12 and item N23 coming from EVAPM3/90 shown i Figure 4, we obtained the contingency table in Figure 5. In the table, 8 percent of the 3rd level students (Grade 9) succeeded both on item D12 and on item N23. The sample used for the computation was formed from about 100 students from 100 different classes, a smaller number than we would have liked, but we are still limited by the cost of acquiring data.

Anyway, in this case, the Chi-square test did not lead us to reject independence. But we will not consider this kind of statistical question in this paper. As we will see the question of the quality of assessments may not be

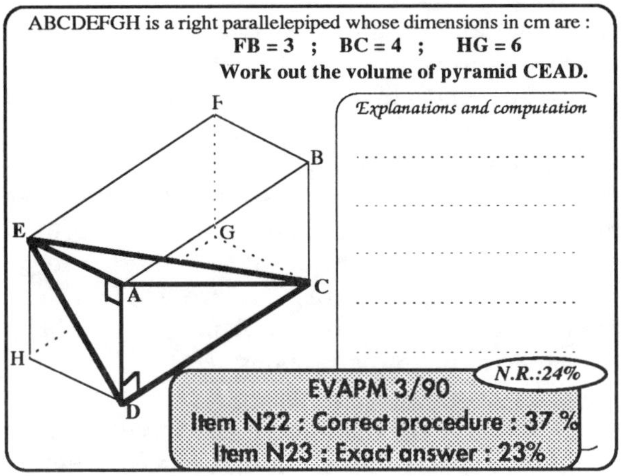

Figure 4

EVAPM 3/90		Item D12		
		1	0	N.R.
Item N23	1	8%	12%	3%
	0	21%	18%	2%
	N.R.	11%	20%	5%

Figure 5

covered in statistical illustrations.

4. SOME OF THE DIFFICULTIES ENCOUNTERED IN ASSESSMENT

When our aim is to be able to say something reliable about individual or collective knowledge, we understand that this is far from easy. Several previous studies have helped us to identify the difficulties.

Here are some of them:

- To change the position of a given item in a particular test may modify significantly the rates of correct answers. So may the time allotted for the test, or the general length of the test in which the question occurs (by simple suppression or addition of other items).
- The operationalization, i.e., actual form taken by the question, chosen for checking a given skill is sensitive to such a degree that the change of only one word (even an article) can produce strong differences in the results. So, what can be said when the question is "dressed" differently?
- The formulation of the questions asked may convey wrong ideas about the knowledge of concern. These errors may be due either to a lack of mastery on the part of the test designer in the domain investigated, or to his desire to make questions simpler. In the two cases, the wrong conceptions of students will not be detected, and the validity of the operationalization used will not be assured.

- The formulation of questions may carry so much weight that it is implicit that a correct answer is more a sign of similarity in the ways of thinking of the questioner and the answerer than an indication of the knowledge of the answerer.
- The information gathered depends, to a large extent, on the form of questioning. For an a priori task analysis according to the form a question takes changes: open, open-ended, closed, or multiple-choice, the same student will not give the same answers. We have much evidence that the behavior of a given student on one form of test cannot be predicted on the basis of his results on another form. Only a posteriori study of the procedures used by students can explain such a difference. Nevertheless, the problem of determining what a student's competencies might be remains.

The "objectives movement" has brought the idea of objective-centered assessments with the educational scene. We have explored that movement for 15 years. This exploration allowed us to break with norm-referenced testing and provided us with technical devices which proved to be very useful. However, lots of problems remain.

5. SOME LIMITS OF ASSESSMENTS FOCUSING ON OBJECTIVES

Objectives often describe "how-to's" which may or may not be associated with real knowledge. For example, let us consider the following objective:

Be able to solve an equation of the type $ax+b=cx+d$.

We have observed children who are able to solve Question 1 but not Question 2:

Question 1: Solve the equation $7x-3=13x+15$.

Question 2: Is 10 a solution to the equation $7x-3=13x+15$?

One can imagine that teaching consists of stressing the mastery of the how-to's to the detriment of meaning. It is, in fact, possible to associate similar examples with most objectives, and we are well aware that we are changing registers (but not really changing levels) in moving from Question 1 to Question 2. What we want to say is simply that, if one is not careful, the technology of objectives may facilitate the learning of algorithms to the detriment of understanding.

In the meaning we give to the word knowledge, we must be clear that a "how-to" or a "know-how-to-say-it" that is not linked to a network of meaning is only a conditional response and cannot even be considered as an element of knowledge.

Here is another objective:

At the end of sixth Grade, a student should be able to make the necessary measurements and calculate the area of a triangle drawn on a piece of paper and given to him.

We cannot help but be upset by the fact that, according to whether the triangle used was one or the other of the two reproduced below (reduced in scale, as are all the figures in this paper), we observed that either half the children attained the objective, or only a fifth of them attained it (se Figure 6).

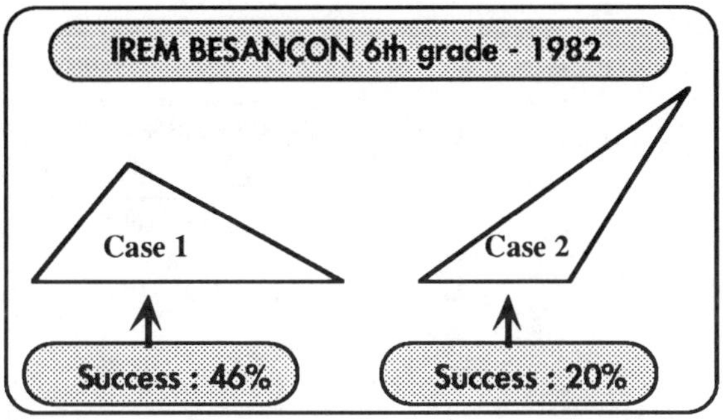

Figure 6

Thus, the fact that the student knows or does not know depends, to a large extent, on the operationalization chosen. Let us have another look at the example of the area of the triangle. Nothing allows us to be sure that a student who succeeds in calculating the area of a triangle in the two cases above will also be able to measure a triangular field and calculate its area (I mean a real field in the landscape). Knowledge always appears in some particular situation and cannot necessarily be mobilized in other situations. In our opinion, a particular piece of knowledge must be situated with respect to the class of problems for which it constitutes an appropriate solution. Here is another example that concerns the ability to draw the bisector of a given angle at the end of 6th Grade (se Figure 7).

Which students attain the objective? One could declare the objective to be attained only by those students who succeed in all three cases but, on the one hand, it would be easy to find a fourth case that would disqualify a good part of the students; yet on the other hand, one can only wonder why the knowledge of the others should be ignored.

By examining the procedures used by the students, it is easy to explain the difference observed between the results obtained with item A18 and

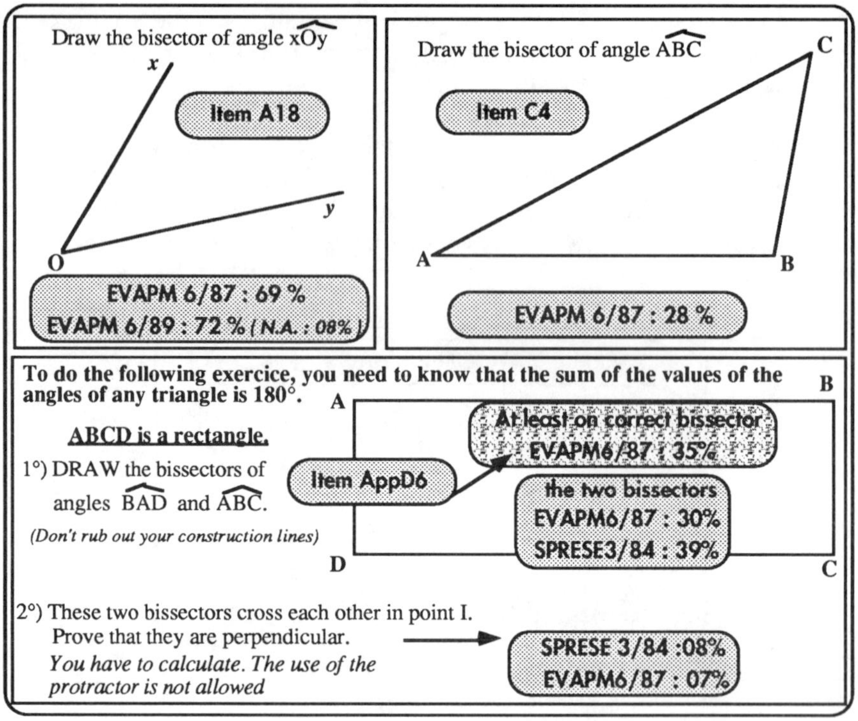

Figure 7

C4. It is only on C4 that students may draw the height or the median from point A. The drawing of the median appears more frequently. Surely, students do not see the contradiction. For them, the median splits the angle in two equal parts and, using a ruler (to place the middle of segment [BC]) is easier than using a compass.

The contingency tables in Figure 8 show that AppD6 implies A18, but 60 percent of the students who succeeded in AppD6 failed in C4, while 30 percent of the students who succeeded in C4 failed in AppD6.

Statistically C4 and AppD6 are not independent, and it is the success in C4 which allows us to predict the succes in AppD6 the best. In this text, we do not have enough room to develop statistical arguments. Let us simply say that very often it is the behavior of 20 percent to 30 percent of the students that is predicted badly by global statistics. But, in one class group, 20–30 percent represents from 6 to 12 students, perhaps exactly those who should be taken into account in a particular way.

Hereafter, (Figures 9 and 10) we show questions asked at 6th Grade level, concerning axial symmetry. Contingency tables are presented when

EVAPM 6/87		Item A18		
		1	0	
Item AppD6	1	38%	00%	
	0	31%	31%	

EVAPM 6/87		Item C4		
		1	0	
Item AppD6	1	17%	26%	
	0	07%	50%	

Figure 8

some students passed two of the questions (that is not always the case — for instance, no student passed both item A19 and item C18). The results confirm strongly what is already known about the influence of some didactical variable, such as the direction of the line of symmetry (se Grenier, 1986).

The simplest figures obtain the lowest results; for the students observed, the "conservation" of the direction appears more often when the direction is a "horizontal" one. Constructing the image of a given triangle about a line is much easier than constructing a segment of a line. A more complex figure may be considered as easier than a less complex one, provided the line of symmetry is "vertical".

But what about competency? When can we say that "a student is able to draw the image of a segment line in an axial symmetry"? Any of the possible successes with each of these five questions appears to be a condition needed to succeed anywhere else. The same is true concerning failures. Meanwhile some successes appear as better predictors of other success than others. For instance, a student who succeeds on C18 is more likely to succeed on AppB16 than a student who succeeds on AppB16 is on C18 (78 percent against 43 percent), at least with the sample observed.

What precedes are only a few examples. Our studies allow us to make such comparisons in many domains. Until now we have said nothing about taxonomies and task analysis methods. Actually we generally use such grids as tools, and we find them useful, but they also appear to be of little help in solving assessment concerns.

5. THE STUDENTS' BEHAVIOR RESPECTS NEITHER THE TAXONOMIES OF OBJECTIVES NOR THE A PRIORI ANALYSIS OF DIFFICULTIES

The use of taxonomies assures a certain variety in the level of difficulty of the questions asked. On the condition that one remains inside a single area of knowledge, that one takes into account teaching practices, and that the analysis of the task confirms the hierarchy of the questions, the aggregated

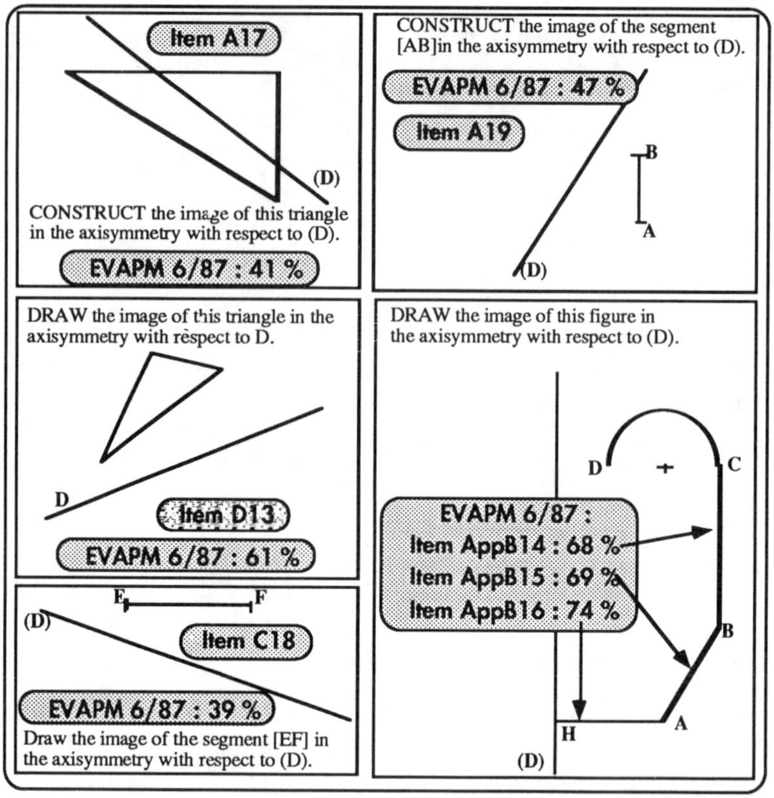

Figure 9

EVAPM 6/87		Item A17		
		1	0	N.R.
Item A19	1	30%	17%	
	0	11%	42%	

EVAPM 6/87		Item A19		
		1	0	N.R.
Item AppB14	1	36%	29%	
	0	11%	24%	
	N.R			

EVAPM 6/87		Item C18		
		1	0	N.R
Item AppB16	1	32%	43%	
	0	09%	16%	
	N.R.			

EVAPM 6/87		Item D13		
		1	0	N.R
Whole AppB	1	38%	25%	
	0	15%	21%	
	N.R.			

Figure 10

results usually conform to taxonomies. It is not all the same at the individual level.

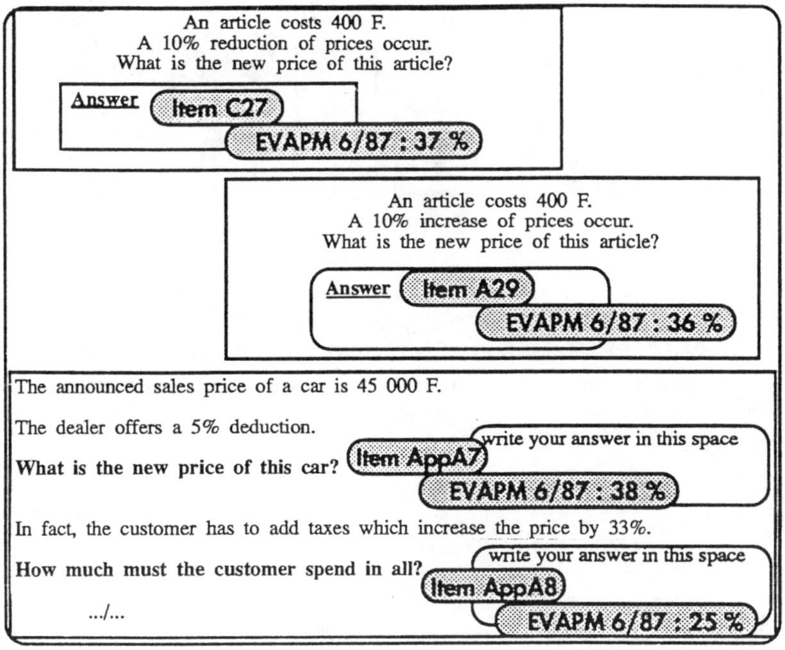

Figure 11

Frequently a pupil succeeds on a Y question and fails on an X question, although the analysis of the task tends to subordinate Y to X.

For example, question AppA7 and C27 were submitted to pupils at the end of 6th Grade. Task AppA7 is more complex, and taking into account the teaching given, is on a higher level of the taxonomy than C27. But 31 percent of the pupils who failed C27 succeeded on AppA7, while 30 percent of those who succeeded on C27 failed on AppA7. Contingency tables in Figure 12 show, in particular, that 35 percent of the students who succeeded in AppA7 failed in C27!

The first solution which comes to mind is to limit the observed effects by producing objectives accompanied by their operational definitions, which will then be considered as unique. That is what we did for a long time and it has lots of advantages: easy communication between teachers and students, non-equivocity, and the possibility of deciding what can be required from students at each school level. However, this method, in a way, declares that "Intelligence is what is measured by ... intelligence tests"; it would be hazardous and worrysome to declare: "Knowledge about symmetry is what is measured by such and such a test on symmetry". We have seen that changing some questions, or even modifying their order or

EVAPM 6/87		Item A7		
		1	0	N.R
Item AppA8	1	24%	01%	
	0	14%	61%	

EVAPM 6/87		Item A29		
		1	0	N.R
Item AppA7	1	20%	10%	
	0	18%	52%	
	N.R			

EVAPM 6/87		Item C27		
		1	0	N.R
Item AppA7	1	28%	13%	
	0	12%	47%	
	N.R.			

EVAPM 6/87		Item C27		
		1	0	N.R
ItemAppA8	1	22%	07%	
	0	18%	53%	
	N.R.			

Figure 12

the time given to take a test is sufficient to turn knowledge to ignorance and vice versa. This problem, the reliability of an assessment, can be extended to the set of tests produced to evaluate a particular kind of knowledge.

6. LEARNING AS A TIME EVOLVING PROCESS

Is a 20 percent success rate always equal to another 20 percent rate? We could easily produce two questions, Question A and Question B, which at 6th Grade level, for instance, could obtain the same success rate (20 percent) but which four years later (at 9th Grade level) would obtain an 80 percent rate for Question A but a 20 percent rate for Question B. So when examining a question, we also need to examine its future, at least under current and known teaching practices.

In the EVAPM studies we tried to follow this future. Figure 3, Figure 4, and some others in this paper show such a continuous observation. Figure 12B shows another question that was used at several levels. Questions documented in this way could be very useful in the design of assessments. At the moment we at the IREM in Besançon are developing a computerized data base to store questions and such documentation.

7. IMPORTANCE OF THE SURROUNDINGS OF THE QUESTIONS ASKED

Most of the APMEP (1987–91) tests are "puzzled", i.e., they are made with the questions taken from different areas, so that the influence of items on each other is reduced. Nevertheless, in a few cases, "homogenous" tests were used (i.e., regrouped questions from the same area). It is hardly surprising that in the case of homogenous tests the results obtained were much better. This again leads us to reconsider the question of competencies. For example (see Figure 12C), on one question the observed difference was important. When this question was included in a "puzzled"

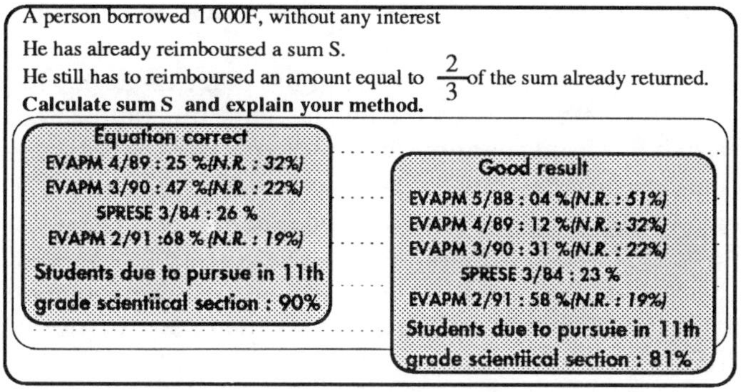

Figure 12B

test, the rate of success was 29 percent. When it was included in a homogenous test, the rate was 51 percent.

EVAPM6/87- D17

John buys a radio alarm clock costing 240 F.
The shopkeeper gives him a reduction of 60 F.

Express this reduction as a percentage

Answer :

Figure 12C

8. INFLUENCE OF STUDENT'S INVOLVEMENT ON THE RESULTS OBSERVED

In our assessments, teachers were allowed to use or not to use the students' results for their own evaluation. Our only requirement was that the students (and we) be warned beforehand. About one third of the teachers seized this opportunity. At each level, the results show a significant difference between the two sub-populations. On an average one may estimate a five percent increase in the scores obtained when students are in a "non-free" assessment; for some questions the difference may read as much as 15 percent. Actually, for a given question, the difference between the rates of correct answers in the two groups, and the rates of non-answers are linked to the effort needed to solve the question. It is more a matter of the personal involvement required than of the difficulty of the question, and the difference appears more often on performance tasks when compared to open-ended or multiple-choice items.

The previous observation confirms our determination to organize assessment in different affective contexts and to compare the results obtained in such different situations before making any generalizations. The different contexts may be those of free or non-free assessment, individual or collective settings, paper-and-pencil, or oral.

With the EVAPM studies we were soon confronted with a question about the discrepancy existing between the official requirements and the apparent knowledge shown by the students. This led to our asking the following question:

9. SHOULD ASSESSMENTS BE TAILORED TO STANDARDS, OR OFFICIAL REQUIREMENTS, OR TO ACTUAL STUDENTS' ABILITIES?

In the APMEP study, the tests were built in such a way that the entire syllabus was covered. As far as "required proficiency" is concerned, we observed that less than half of the students showed mastery on more than half the questions they were asked. Fortunately, we took the precaution of linking our study to the former ones and, in all cases, the results observed were as good or better than those of the last ten years. So, it is not possible to state that there is a drop in students' level of performance. At the very time when the authors of the syllabus thought that they had reduced requirements to those that should be attained by 80 percent of the students, we found that less than five percent of the corresponding questions met their condition. Reducing the requirements to those questions really mastered would certainly have a main effect of reducing teaching to a few meaningless tricks. It is actually possible to enact official requirements which can be put into questions. This point is developed in Bodin (1989).

10. STATISTICAL DEPENDENCIES AND GENERALIZABILITY

Usually, a given question or a given test is valuable, not for what it allows us to say about it, but for what it allows us to say with confidence about a larger set of questions which were not asked. This is indeed another way to tackle the problem of competencies. Some of our tests, called "required", aimed to cover closely the official list of required proficiencies, while others, called "complementary", sought deeper skills. The obvious assumption that "he who can do a lot can also do a little" is often shown to be false.

Here is a scatter plot (Figure 13) of the results of a sample of 1,600 students out of 1,600 different classes (EVAPM5/88). It shows the relationship between scores on the required proficiencies (EX-TOTAL) and the deeper skills (COMP-TOTAL). Indeed the correlation is high ($r=.705$), but we can estimate at 25 percent the proportion of students for whom the results obtained are quite contradictory.

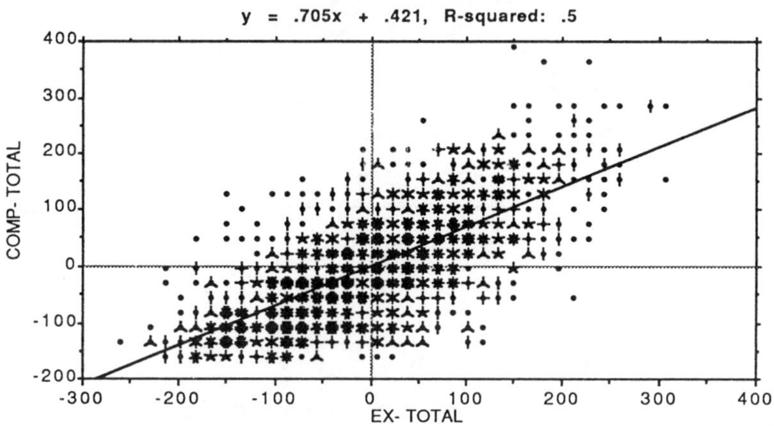

Figure 13

In fact, the difference is even more impressive if we separate the set of questions into three levels of difficulty:

Easy: $R>60\%$ (more than 60% success)
Mean: $30\%\leq R\leq 60\%$
Difficult: $R<30\%$

In this case we obtain a very low correlation between the scores achieved by the same students on the two subset of items, as shown in Figure 14.

We found the same result when analyzing lots of tests:

- Students who have high results (above average) on the set of easy questions may have either high or low results (below average) on the set of difficult questions.
- Students who have low results on the set of easy questions may also have either high results or low results in the set of difficult questions, but this occurs to a lesser extent than for the previous case, at least as long as what distinguishes the two sets of questions is only the

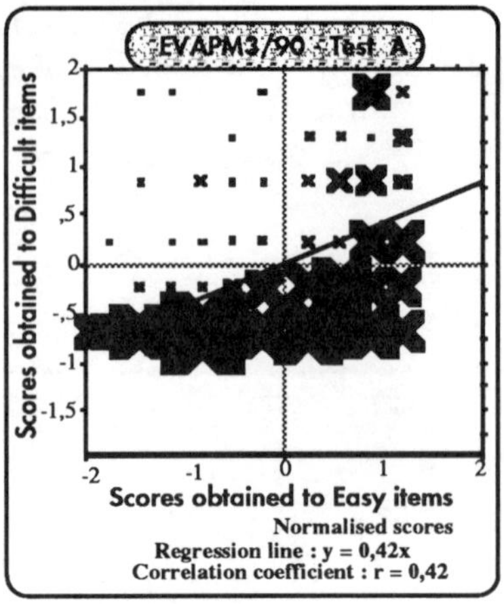

Figure 14

difficulty (measured by the scores obtained) and no other factors such as their taxonomical level.

- Generally, but with note-worthy exceptions, a student who succeeded on the set of easy questions is more likely to succeed on the set of difficult ones.

Figure 15 shows a cross tabulation of results that were obtained on required questions (official requirements) with results obtained on a test centered on problem solving processes. For this later test, scores were not normalized, but this does not inconvenience our purpose. Limiting or stressing the assessment of easy questions, required questions, or lowest taxonomical level questions is sure to prevent us from taking account of students' existing competencies.

Instead of considering the previous test concerning problem solving, we considered only a performance task from this test that asked for creativity and discovery; the correlation coefficient fell to 0.15 and an independence test did not allow us to think of the two kinds of results as dependent.

To summarize, links are weaker than expected between students' performances when they are obtained in different contexts. Every time we tried to find a hierarchy of students' mathematical abilities, we had to conclude that such an order does not exist. We reached the same conclusion when crossing the results of students in different areas of mathematics. For instance, when crossing the scores obtained in numerical questions with those obtained in geometrical ones the correlation coefficient was usually around 0.6; while it is statistically significant, it hides the fact that, for half of the students, the link between their results in different fields is very

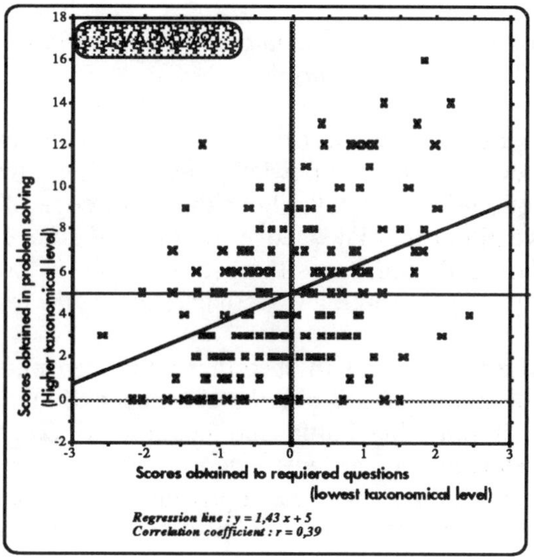

Figure 15

weak.

11. EVALUATION OF KNOWLEDGE LACKS RELIABILITY

There are certainly tests that are more reliable than others, but it is clear that when one gives the same test to the same student at a later time, one never knows what one will obtain. Even without new learning, time has done its work: Student knowledge has been modified or restructured; formal knowledge has been forgotten whilst new meanings have appeared — in short, even if the test is the same, students have changed. It is not even true to say that the test is the same, since a test is not simply a printed paper. It contains a certain number of situational variables which, with teachers being only more or less aware of them, exert an important influence.

In reality, a test is only one operationalization out of a possible set, big or small. We have just pointed out, by talking about the same test, that consistency of results are not assured. If, as well, the choice of objectives or operationalizations (setting of questions) are modified, or if teaching is modified, the behavior of pupils becomes totally unpredictable. What is not reliable, then, is the judgement that one can make about the knowledge of a pupil by using one or several particular tests.

12. WHICH PERSPECTIVES?

Focusing More on the Meaning of the Information Gathered Than on Its Value

Generally, theories and discourses about assessment avoid taking into account the specificity of the knowledge in question. However, when the objective of assessment is to talk about the knowledge of an individual or a group, and we admit that knowledge can only be inferred on the basis of the information gathered, and that (moreover) it possesses its own dynamics, we quickly become aware that ready-made recipes are insufficient.

In France, the present growth of studies on the learning of mathematics (the didactics of mathematics), and the concepts and methods that are put forward make it both inevitable and possible to take knowledge into account in a relevant way, at least when knowledge is a part of the real objectives. Starting from a technical approach to assessment, we have been influenced strongly by the work of numerous researchers in the field "processes of mathematics learning". We will mention only three of them for the additional light they have given us.

Vergnaud, G. (1984, 1990) for his work in *conceptual fields* and cognitive structures, representations, and theorems in action (operational invariants). Little by little he led us from a behaviorist vision to a constructivist one. For those who take this path, a certain number of "truths" about knowledge rapidly disintegrate.

Brousseau, G. (1986) for his work on the modeling of learning situations. The possibility he has given us of identifying games, players, strategies ... in learning situations, the importance he gives to the *didactical contract* as the (implicit) rule of the game and above all, the importance of the breakdowns of this contract, have helped us considerably to clarify our conceptions of assessment.

Chevallard, Y. (1985) who has shown us to what degree "the fact of assessment does not only have contingent existence, is not only a necessary evil that one could ignore, but is in fact one of the determining aspects of the learning process — which controls and regulates both the behavior of the teacher and the learning of the students", showing that in the classroom (and in our opinion, throughout the whole school system) the process of assessment depends more on compromise than on expertise. He has confirmed our belief that a too exclusive attention to knowledge could lead to a kind of reductionism.

The didactical approach led us to put aside the idea of measurement, at least as long as we do not know what is to be measured, and even if there is something to measure.

Assessments often focus on the rate of good answers obtained, but wrong answers and non-answers are also worthy of being examined. Above all, the procedures used by students when answering questions should be studied. Actually I think it is most important that we move from a quantitative view of assessment to a qualitative one. Meanwhile, in this chapter we have stressed the lack of reliability of the information gathered, at least when the information is isolated. As in a criminal inquiry, we must cross and cross check again the information gathered, to make meaning appear.

We are still lacking a theoretical framework that would join the pieces of meaningful information we collected. The Vergnaud (1990) conceptual field theory is very promising for this purpose. It allows us to model, if not knowledge as a whole, parts of knowledge linked with a set of situations. With this theory, the prospect of *concept-referenced testing* becomes more thinkable. The term concept-referenced testing appeared in the literature a few years ago, likely first introduced by Horne (1987), but in my opinion with a questionable theoretical background regarding categories of learning.

The data analysis methods available for structuring conceptual fields have been enriched for a few years by Gras' statistical and didactical work concerning the implication theory. To show the work I quote Gras and Larher (1991):

"If the problem of the simultaneity of two events A and B finds part of its solution in the symmetrical study of correlation or similarity, the question of inference (if A then B) has on the other hand to undergo the examining of an asymmetrical relationship. In relation with a psychological problem of complexity, R. Gras brought in his thesis a contribution that made possible a number of applications of this type of relationship in research in psychological genetics or in mathematical didactics. But the variables taken into account in his research are restricted to binary ones, presence-absence of a feature in a given individual. The proposal here [i.e. the latest work] is to extend the study of the statistical inference (or quasi-inference) to other types of variables and above all to *classes of such variables*. This extension allows us to build a *tree of oriented classes*."

The method has been applied to some of EVAPM assessments and seems to give interesting results. The work is still in progress and at the moment we may only give a general idea of the way it works.

As a fictitious example, 10 items processed with the implication method are shown in Figure 16. Item 3 is closer to Item 2, for instance, than Item 6 it to Item 1. This means that a student who succeeds on Item 3 is likely to succeed on Item 2, and that this is more likely than a student succeeding on Item 6 is likely to succeed on Item 1. The grouping into classes, as shown in Figure 17, leads us to consider success on class A as a good predictor for class B. The problem is to give meaning to these statistical classes. In a given conceptual field, content analysis of items and an analysis on the procedures used by students usually can be done. So, conceptual fields and implication theory methods might lead to a map of students' knowledge or competencies.

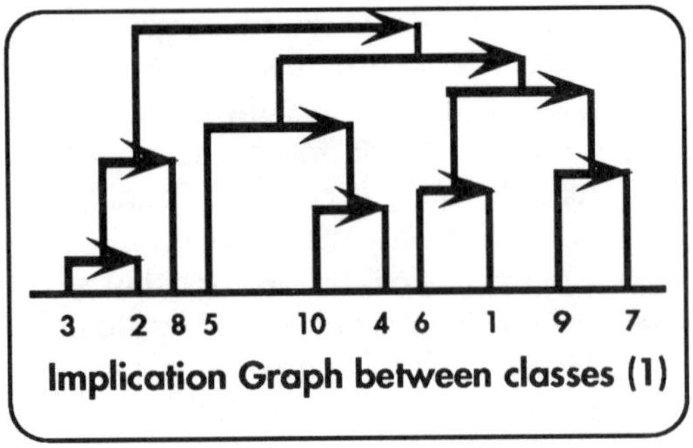

Figure 16

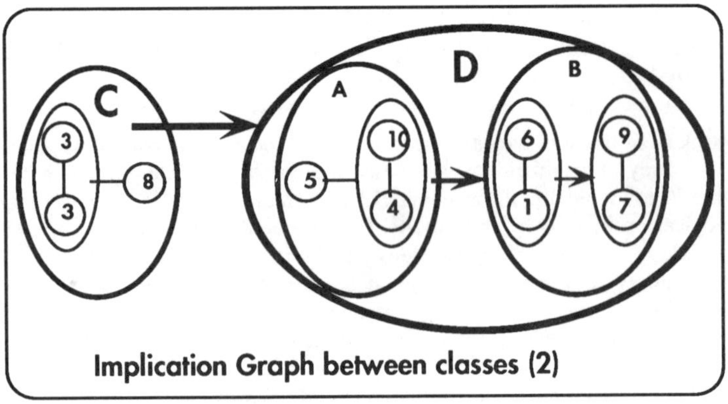

Figure 17

Actually we obtain a dynamic map that is very different, and more useful for assessing purposes, than a static one. It still misses something necessary if we are to have a satisfactory assessment framework. When beginning this chapter, we used the metaphors of position, speed, and acceleration. Here we have good clues about position and tendencies, but not yet about speed. In Figure 17, when a student masters class C it is quite likely that he can also master class D (discrepancies have to be explained), but what happens if he masters class D but not class C? The task is to estimate how far this student is from the mastery of class C. By that I mean what energy the teaching/learning process has to display to ensure the mastery of class C. Surely the Vygotsky (1978) *zone of proximal development* concept may help us to tackle this question.

It would not be true to say that we are much more advanced in this approach, but we can point out how useful it must be to take this question into account, particularly because such usual thoughts concerning assessment prevent us from providing students with additional help. More exactly, help is usually given by organizing the situation from its simplest aspects to its more complex ones, or by decomposing difficulties. What would follow from a Vygotskian approach would be to observe students tackling a whole complex situation and, only afterwards, to observe what additional information they need to solve the question.

To explore conceptual fields, as well as for assessment needs, we think that we need a typology of assessment tasks that account for the meaning a given task may have in an non-artificial mathematical context, e.g., to someone who really uses mathematics either at a real life level or as a mathematician. For this purpose, we are using a very simple typology that proved useful and has the power to realize to what extent assessments usually focus at the objective level of mathematical concepts.

13. A TYPOLOGY OF MEANING

To analyze relevance and validity, we combined the use of the usual taxonomies with what we call a *typology of meaning*. In order to do that, we distinguished several levels at which a concept may be assessed or where it may intervene, without the intention of placing each in a hierarchy. Thus, each question would be situated in relation to the meaning of the knowledge involved.

This typology is described in brief:

The Representation Level

The representation level concerns mental images and physical images of the concept (signifiers). This level brings out the meaning that the student gives to the concept or notion, independently of the techniques he may have to solve the questions where the concept arises, as well as his means of expressing it. For example, to imagine straight lines as infinite is not on the same level as being able to say by rote, "all straight lines are infinite", a learned sentence that does not necessarily have a clear meaning for the person who says it.

Figures 18–19–20 show questions assessing this level. Perhaps we should use the word *perception* instead of representation, but we want to stress that it is the perception of mathematical concepts and not social perceptions. We also have to insist on the fact that representations are not necessarily geometrical or graphical representations. For instance, students may have representations of numbers as forming a discrete set, or of functions as compulsively introduced by an algebraic formula.

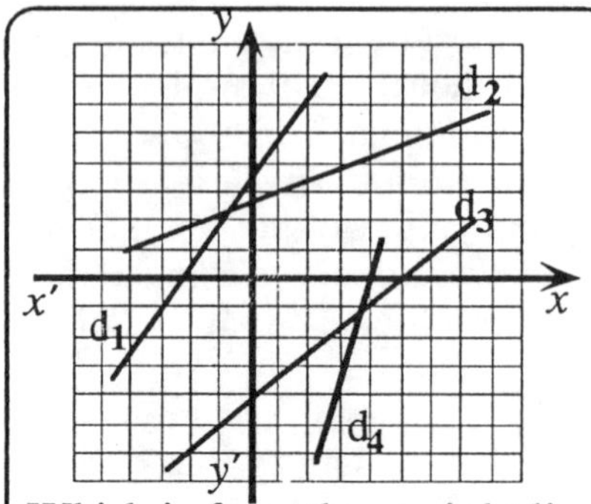

Which is from the straight lines d_1, d_2, d_3, d_4, the one which has the smallest gradiant ?

EVAPM 2/91 : 61 % (N.A. :02%)
Oral test (Without paper and pencil)

Figure 18

Points O, A, B and C are set according the figure herebelow.

CONSTRUCT one point M belonging to the straighy line (OC)

such as : $\frac{OA}{OB} = \frac{OC}{OM}$

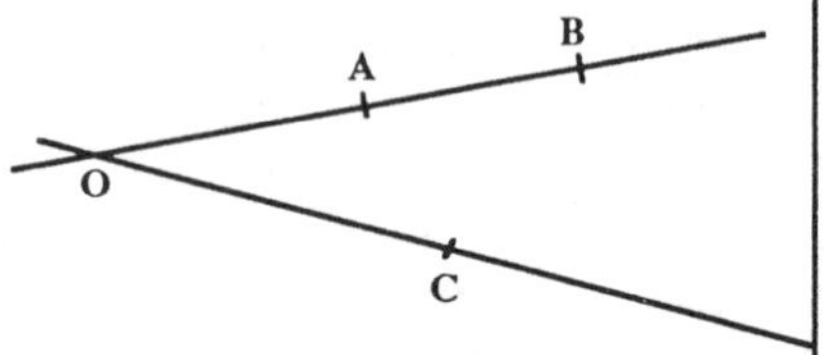

Don't rub out constuction lines

EVAPM 3/90 : 62 % (N.A. : 07%)

Figure 20

I have drunk the third of half a liter of milk.

What fraction of a liter did I drink ?

EVAPM 3/90 : 42 %
Oral test (Without paper and pencil)

Figure 19

The Communication Level

In many cases, a concept is only used to describe or to encode information. Here, the concept intervenes on the communications level. This is, for instance, what happens to the concept of vector when it is stated:

ABCD is a quadrilateral such that: $\overrightarrow{AG}=2\overrightarrow{CD}$

Figure 21 gives example of one question belonging to this level.

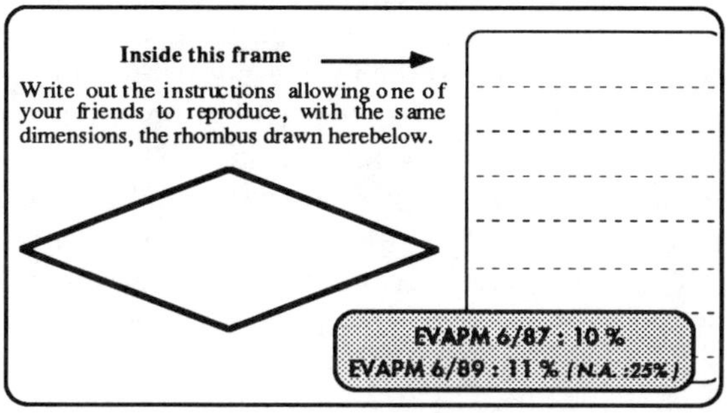

Figure 21

The Object Level

In this case, the task relates to the mathematical object itself without any concrete framework that gives it another type of meaning. For instance, a student can solve a given equation without being able to express the steps he took or justify his results without knowing to which type of problem it is connected, or without being able to use it as a tool in another situation. Unfortunately the school experience of most students is quite exclusively of this kind, not only in assessment but also in teaching practice.

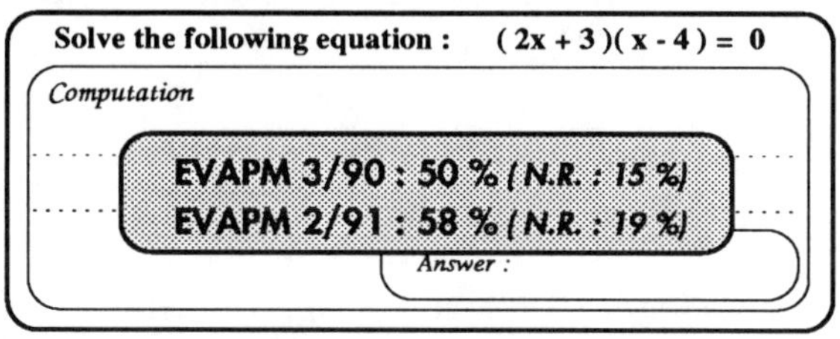

Figure 22

The Tool Level

To quote Douady "By tool, we mean the scientific functioning of the concept in different problems it is able to solve". A concept takes on meaning through its tool characteristics, and also "... to have mathematical knowledge means to be able to use it as explicit well-adapted tools with reference to the problems which gives them meaning". Figure 23 shows a question from this level. Problem solving questions and performance tasks may generally be situated at this level. The various instruments mentioned above provide a means for constructing assessment tools and for facilitating their interpretation. However the useful information provided by a given assessment will vary according to the function attributed to it. For instance, assessment may be formative on the condition that the information gathered sheds a revealing light on certain elements of the student's knowledge and that this information is available and usable by the student, directly or through the teacher. But in fact, there are still other conditions: that the state of the didactical contract allows the assessment and the communication of results to have a formative effect. But this is a another story.

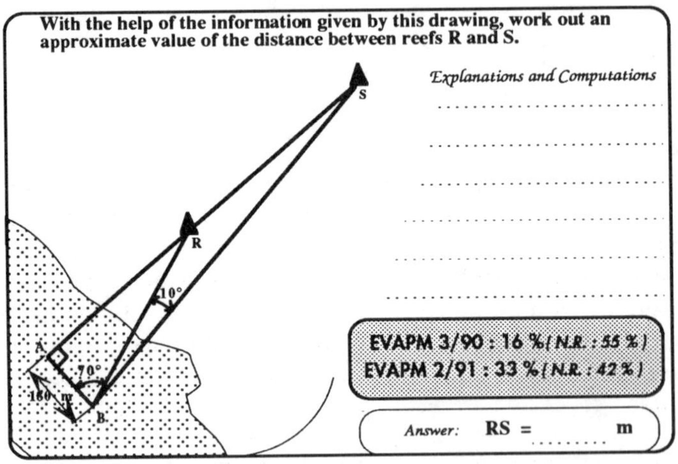

Figure 23

14. BEYOND DISTINCTIONS OF FORMATIVE ASSESSMENT AND SUMMATIVE ASSESSMENT: DIAGNOSTIC ASSESSMENT

Whether for summative or formative purposes, the assessment of knowledge may always lead to a diagnosis and, in general, it should. The difference is simply one of the scale of the area of knowledge that is taken into account. An end-of-the-year decision, for example, the result of summative assessment, although it is a *prognosis*, is also diagnostic. In fact, it is supposed to constitute the best possible solution for improving the knowledge of the student in question. In the same way, the many decisions that a teacher makes (choice of exercises, organization of situations in the class) are supposed, each time, to constitute the best possible solutions for the improvement of students' knowledge.

To be diagnostic, assessment must take differences between students into account. It must be varied, if not individualized. The illusion of justice that consists of treating in the same way all the year long students whose knowledge state is not the same, is an important obstacle to a real formative assessment.

An assessment question must be considered as an indicator of the state of student's knowledge. For this purpose, the study of the procedures followed by the students is essential, as is the study of the statistical dependencies observed through large and repeated use of the question with the reservations mentioned above.

Here is an example of an indicator-question: The Wall. (See Figure 24).

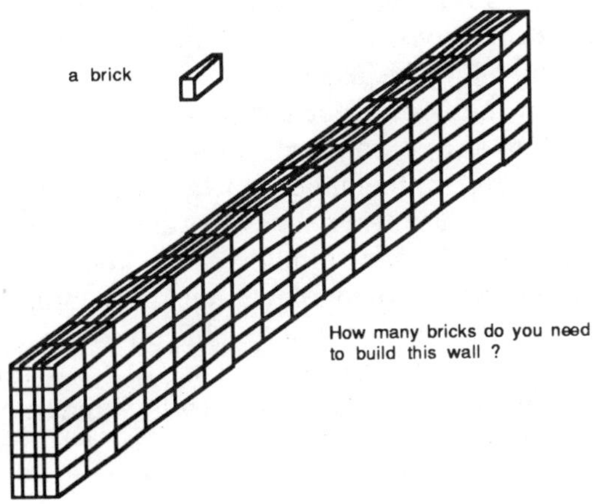

Figure 24

In 7th Grade, students usually solve this question in Figure 24 easily. However, for some of them, a conflict appears between two types of procedures: measurement procedures and counting procedures. In the case of counting, corner bricks have to be counted up only once, while in the case of measurement it seems that corner bricks are counted once in each direction. The time spent solving the question is a more powerful indication than the answer given of the state of the student's knowledge. To place this question into a network that models knowledge as it functions in the conceptual field of "multiplicative structures" would surely be of great interest.

15. CONCLUSION

Most of the work on assessment shows that across school systems, assessment is often done with total confusion as far as its functions and its implications are concerned. Some of the proposed solutions merely suggest entrusting assessment to specialist and improving measurements devices. I think measurements aspects may still be relevant if based on serious *a priori* epistemological and didactical analysis. They may also clarify the underlying hypothesis which limits the area of the validity of the statistical models used and which is more often than not forgotten.

Concerning the delegation of assessment to specialists, we believe that we have shown that as far as mathematics is concerned, such a specialist intervening in a class, should have better training in mathematics than the teacher has, and at the same time, he should know the students of that class better than the teacher does. That is, perhaps fortunately, impossible. But test designers, should they be teachers or not, must tend to become specialists of assessment methods. Teachers should avoid confusing good learning situations and good assessment devices.

We are aiming to develop mathematics teachers who can do a rigorous analysis of the methods they use when assessing and to supply them with indicators that lead to a better understanding of the knowledge of each of their students. For a long time, doctors have not made drugs themselves. The time will come when teachers will not be obliged to make all the tools they need themselves. They may then have a set of tools for accurately diagnosing the state of their students' knowledge, and more efficient learning will be the result.

It is to this end that we are working.

REFERENCES

APMEP (Association des Professeurs de Mathematiques de l'Enseignement Public). (1987–1991). Evaluation du programme de Sixième – Seconde. Paris: Etude coordonnée par Bodin, A, IREM.

Bodin, A.: 1985, 'Problèmes de l'évaluation des savoirs mathématiques', *Petit X*, no **7**/1985, IREM de Grenoble.

Bodin, A.: 1988, 'L'évaluation du savoir mathématique', *Bulletin de l'Association des professeurs de mathématiques*, no **368**/1989, 195–219.

Bodin, A.: 1989, 'Est-il possible? Est-il souhaitable? De spécifier les "compétences" attendues en fin de formation. Le cas de l'enseignement des mathématiques', *Actes de la 41'eme rencontre C.I.E.A.E.M*, Bruxelles, Juillet 1989.

Bodin, A.: 1991, *Replication et compléments, 1991, des évaluations Sixième et Cinquième*, IREM, Université de Besançon.

Brousseau, G.: 1986, 'Fondements et méthodes de la didactique des mathématiques' *Recherches en Didactique des Mathématiques*, **7.2**, Ed la pensée sauvage, Grenoble.

Carpenter, T.P.: 1980, 'National Assessment: A perspective of mathematics achievement in the United States', in *Proceedings of the 5th International Conference of Psychology of Mathematics Education (P.M.E.)*, Univ. of CA, Berkeley, CA.

Chevallard, Y.: 1985: 'Vers une analyse didactique des faits d'évaluation', in *L'évaluation: Approche descriptive ou prescriptive*, actes du colloque ADMEE, DeBoeck Université, Bruxelles.

Chevallard, Y. & Feldmann, S.: 1986, *Pour une analyse didactique de l'évaluation*, IREM d'Aix, Marseille.

Douady, R.: *Jeux de cadres et dialectique outil-objet*, thèse de doctorat, Université de Paris VII.

Gras, R. & Larher, A.: 1991, *L'implication statistique, une nouvelle méthode d'analyse des données*, IRMAR, Université de Rennes.

Gras, R.: 1986, 'Recherches sur l'apprentissage: Analyse des correspondances et méthodes statistiques apparentées', *Cahier du Cirade*, Université du Québec, Montreal.

Grenier, D.: 1988, *Construction et étude du fonctionnement d'un processus d'enseignement sur la symétrie orthogonale en sixième*, thèse-Université Joseph Fourier de Grenoble.

Horne, S. 1987, 'Concept referenced testing', *European Journal of Psychology of Education*, **II**.

Nuttall, D.L.: 1987, 'The validity of assessments', *European Journal of Psychology of Education*, **II**.

Robin, D. & Barrier, E.: 1985, *Enquête internationale sur l'enseignement des mathématiques — Le cas Francais*, Institut National de Recherches Pédagogiques, Paris.

Romberg, T.A.: 1988, 'Evaluation, a coat of many colors',in Robitaille, D. (Ed.) *Evaluation and assessment in mathematics education*, Science and Technology Education Series No.2, UNESCO, Paris.

Romberg, T.A. & Stewart, D.M.: 1988, *The monitoring of school mathematics*, Wisconsin Center for Education Research, Madison, WI.

Ruddock, G.: 1986, 'Mastery learning and diagnostic testing', in *Proceeding of the 10th International Conference of Psychology and Mathematics Education (P.M.E)*, University of London, London.

Stake, R.: 1975, *Evaluating the arts of education: A responsive approach*, Merrill, Columbus, Ohio.

Vergnaud, G.: 1990, 'La théorie des champs conceptuels', *Recherches en Didactique des Mathématiques*, **10** (2.3), Ed la pensée sauvage, Grenoble.

Vergnaud, G.: 1984, 'Interactions sujets situations', in *3ème école d'été de didactique des mathématiques*, Universite de Grenoble 1: IMAG.

Vygotsky, L.S.: 1978, *Mind in society. The development of higher psychological processes*, Harvard University Press, Cambridge.

Antoine Bodin
Institute of Research on Mathematical Teaching (IREM)
University of Besançon,
France

STIEG MELLIN-OLSEN

A CRITICAL VIEW OF ASSESSMENT IN MATHEMATICS EDUCATION: WHERE IS THE STUDENT AS A SUBJECT?

1. INTRODUCTION: THE CONTRADICTION OF ASSESSMENT

This chapter consist of two parts. In the first part a specific problem related to current assessment practice is discussed. This problem concerns the prevalent status of *the learner as an object* in the process of assessment. In the second part, practical hermeneutics is examined as a possible tool to promote *the learner as a subject* in the process. A case study from teacher education demonstrates the processes.

The learning of the student is to be assessed. During the learning process, the student has the status of subject. During the process of assessing he will usually have status of object. An outsider will, in some way, estimate the quantity and quality of his knowledge as a result of the teaching he experiences.

This dialectical contradiction becomes clear in the goal-directed activity theory as developed in the tradition of Vygotsky[1] and Leont'ev. This theory is based on a hypothesis that all learning is goal directed. Learning thus takes place in the context of a goal. The educationist who works within this tradition has to establish a goal for the learning process. The student who has the status of subject within this tradition, has to share this goal with the educator. It is thus the student, as a subject and as an active learner, who has a goal for the learning process, and who learns in the context of this goal.

Within this tradition, however, the origin of the goal is rarely questioned. Did it originate with the curriculum? With the textbook? or With the teacher? We do not see the student come up with goals, nor do the teacher and students develop goals through dialogue, within this tradition. It is more or less taken for granted that the student shares the goals of the teacher (or the teacher as an agent of a curriculum), or that the teacher manages to get the student to take on a given goal.

The inappropriateness of such an approach becomes clear when the educationist looks closely at the learning process. The phenomenon of pretending by a student to follow a specific goal, or mock cooperation with the teacher, has been observed by many educationists, either by participant

M. Niss (ed.), Investigations into Assessment in Mathematics Education, 143–156.

observation in classrooms or through anthropological fieldwork with students.

Instrumentalism, *chameleon behavior* and *mock cooperation* are some of the metaphors which have been used for this behavior (Mellin-Olsen, 1981, 1987). Chameleon behavior may be of special interest here, as the chameleon changes its color according to the environment in order to make itself invisible. If chameleon behavior is allowed to grow, resistance and avoidance as described by Freire[2] (1970a, 1970b) and Giroux (1983) are inevitable.

So what do we assess when we assess the student's learning? Do we assess the student as an object, by estimating how he handles knowledge in various situations? Or do we relate our observations to how he understands the goals set for him, and the learning process that follows? In other words, how is it possible to assess the meaning (subjective) the student gives to knowledge, as distinct from an objective assessment of his knowledge?

2. HOW CAN WE DISCUSS POSSESSION OF GOALS?

A few metaphors have been developed to provide some insight into the tensions which the contradiction of assessment implies. So far these metaphors have not been developed into concepts that may be operationalized for a theory serving assessment practice.

Guy Brousseau has introduced the notion of *the didactical contract* (Brousseau, 1987). The teacher and the students negotiate about some goal of a learning process. The result of this negotiation is the contract. The students and the teacher commit themselves to be loyal to the contract during the process of teaching and learning. The significant moment in this process occurs when students break the contract, i.e. when they handle the knowledge in a way which was not thought of during the negotiations.

The paradox becomes clear: All possible ways to handle knowledge cannot be foreseen at the moment of negotiation. If the students' possible uses of knowledge are restricted to those being negotiated, then their status as learners is that of objects. If they achieve status as subjects, their teacher must risk letting them do something with knowledge other than what was expected. This latter situation prepares the ground for another metaphor, the teacher's *release of knowledge* which makes it possible for the student *to treat knowledge in his own way*.

This can be done at different levels. Brousseau gives examples of didactical contracts related to mathematical games, which permit the invention of new rules, which, in turn, may lead to new and unexpected mathematical knowledge. The contract is thus related to a specific set of rules for the game. The transformation of the contract is related to a transformation of the rules of the game.

Mathematical model building constitutes another level. A social phenomenon is to be described by a mathematical model. It is taken for granted by students and teachers that a certain set of mathematical relations are to be used as tools for the construction of the model. During the construction process the student discovers the significance of reconstructing the model, by, say, redefining its premises. In this case the student treats knowledge in his own way according to the redefinition of the premises.

Project work constitutes another level. A goal is set for a project by teacher and students. It may be a project about pollution in the environment, about traffic density, or about nutrition. Teacher and students negotiate a strategy. During project work, the students decide to do the project in a way that differs form what was planned by transforming its goal. Rather than investigating public statistics, as the teacher had suggested, they decide to carry out an interview survey. Rather than going for measures of increase in traffic density they may decide to go for statistics about the availability of public transportation. In these cases, the teachers' motivation and their ability to release the didactical situation become significant.

Brousseau introduces another French intellectual, Denis Diderot (1713–1774)[3]. In particular he uses Diderot's *Paradoxe de la comédien* as a metaphor for a healthy relationship between learner and knowledge. The paradox goes like this: In order to appear trustworthy in relation to his audience, the actor must 1) be able to use all his life experience to live the role which is to be performed, and 2) keep an emotional distance from the role. The actor has to live the emotions of the person to be played yet keep an emotional distance. The actor who will play a furious person has to appear furious on stage. Thus, in order to convincingly portray to the audience the ferocity of the character, the actor has to keep an emotional distance to it. The trustworthiness of the actor is linked to his relationship with his audience: The audience is to interpret his acting. It is not to be programmed for a specific interpretation. If the actor does not manage to control his emotions by keeping them at a distance, the audience will miss this opportunity of interpretation. Confusion will arise: Does it interpret how the actor impersonates the emotions or the actor himself?

Brousseau uses this paradox as a metaphor about the relationship between knowledge and the individual. Only by keeping knowledge at a distance is it possible for the individual to control knowledge. If knowledge becomes an organic part of the individual in such a way that the individual *is* his knowledge, he is no longer able to handle knowledge in a controlled way.

I use the paradox as a metaphor of the student-teacher relationship, the teacher as the actor and the students as the audience. The teacher is to convey knowledge. The students are to interpret this knowledge and its possibilities. If the teacher does not manage to control her relationship to the knowledge, she will convey knowledge and herself as a unity. The message to the students will be — if you take the knowledge, you have to take me at the same time. The teacher's role, as experienced by her student, is not that of a conveyor of knowledge. Her role is extended to a larger unit, consisting both of herself as an individual and the knowledge she transmits.

In this latter case it is hard to see how knowledge can be released so that the students can acquire knowledge in their own way. The above does not indicate that the teacher should not be inspired or engaged when teaching, trying to set up ideals, etc. for her students. It indicates, however, that she must still try to keep knowledge at a distance, so that her students are offered possibilities to interpret the significance of knowledge independently.

I have introduced a few metaphors to give some meaning to phenomena such as instrumentalism and chameleon behavior in classrooms. These metaphors invite dialectical interpretations: Knowledge may be released to a certain degree, but not fully. Students can acquire knowledge in their own way for a while, but sooner or later they have to be brought back to a major track again. We are, of course, hinting here at the individual's socialization into his society. Society, usually represented by the State, has as a major task the reproduction of the society. Until a revolution takes place, society is reproduced according to certain values, ideologies, and power structures. During the last 10–15 years, social scientists have found it profitable to stress the *productive* dimension of socialization. Scientists such as Henri Giroux (1983), Anthony Giddens (1977, 1979), and Paul Willis (1977, 1981) have seen the significance of constructing a model of man-in-society which stresses man's production of the social reproduction.

It is claimed that reproduction includes a productive dimension. An ideology is not only *re*produced by a school; it is also produced by those who work in school. Just by working in school or in another institution, we make the institution possible. At the same time as following some old track, we not only reproduce the track, but we also produces its mere existence by using it[4]. In this way we can also say that knowledge is reproduced by the individuals in their production of it.

In this production/reproduction polarity we find a tension of socialization. The new generation can be productive with its uses of knowledge, but not too productive. Our students should be creative with their uses of knowledge, but not too creative. They should bring knowledge new ways, but not too many new ways. Knowledge should be released, but not too much.

In order to restore the student as a subject, it is necessary to keep this tension, as a dialectical contradiction, alive. The student is a subject who may acquire knowledge in his own way. As a member of society, under a certain control of the State, the freedom to acquire knowledge in new ways is restricted. One of the characteristics of the authoritarian or nondemocratic State is that it does not promote the uses of knowledge in new ways in its schools.

So what do we assess? How do we assess? The most difficult task is certainly to assess the subjective uses of knowledge, which may include the breaking of didactical contracts and may challenge the tolerance of the State. How can the educator assess the quality of students' knowledge when they acquire knowledge in their own ways, as long as the educator does not know about these ways? And, to what extent is the educator in a professional position where such forms of assessment are legitimate and desirable?

3. PRACTICAL HERMENEUTICS

I suggest *practical hermeneutics* as a tool for making the student the subject of assessment. Before bringing in some characteristics of this method I shall give a demonstration. This demonstration pictures assessment in a course in teacher education. Such an assessment is clearly of a different type than assessment in a, say, comprehensive primary or secondary school. The principles of practical hermeneutics as practised here are, however, the same as they will be in a classroom at a lower level of education. Listening, observing, interpreting, returning interpretations to the student, and so on, are all assessment procedures that require skills of the teacher, whatever level the teacher works at. As it is the student as a subject who is to be promoted by assessment, interpretations are to be understood as "interpretations of the student's understanding of issues related to her or his own learning".

Assessment of Postgraduate Teacher Education by Hermeneutics. A Case Study.

I shall analyze interview texts derived form qualitative research interviews with Ivar. There exist two interview texts derived form conversations with him about how he understands the course he follows. These interviews are two out of forty made to assess teachers' didactical understanding of mathematics in relation to the courses we offer our students. Ivar was a student in a postgraduate course in mathematics education at Bergen College of Education. The class met one evening each week over an academic year. The content of the course was influenced both by goal-directed activity theory and by radical constructivism. The students had to develop teaching projects reflecting the ideas of these paradigms. Half of the public/formal assessment was based on project work carried out in a school class, the other half on a traditional exam with problem solving. As many of the students were experienced teachers, there were usually hot discussions about methods and goals of mathematics education in the class.

Clearly it will not be possible to evaluate the course in general by analyzing the understanding which one particular student has about it. The case study offered here is an exemplar, in the sense that it demonstrates the uses and possibilities of modern hermeneutics for assessment.

Ivar, a primary school teacher, 32, is in a half-year course in mathematics to gain further formal competence as a teacher. He was interviewed both at the beginning of the course and towards the end of it. In between he gave his class of 8-year olds an addition problem before they had learned the appropriate addition algorithm. He followed up by interviewing four kids about their methods when subtracting.

We first meet Ivar in the second week of the course. Both interviews flow nicely. Ivar does most of the talking. Our conversation is marked by his insecurity. Expressions such as "I am not sure" and "It is a dilemma" are used frequently.

"I like teaching math. My impression is that those I teach are happy with the subject. ... But I do not know whether I manage to get my pupils to look experimentally on the subject". "This balance between learning techniques and developing thinking".

The word "experimentally" comes slowly from him. I take the use of it as a sign of how he already has conceived the expectations of the course. As an interviewer I soon develop two possible working hypotheses for interpretation:

H1. He thinks his present didactical handling of knowledge is appropriate. The course he is offered is wrong for him. This insecurity is a way of surviving in the interview situation as he is confronted by a teacher educator.

H2. He is not satisfied with his didactical handling of knowledge. The course he is offered seems to be the right one. This insecurity is related to what the course may lead to.

More textual material from the first interview:

"The most important to me has been to transmit some of my own joy over mathematics to the pupils" ... "We associated, reflected, and found new tasks and so on" ... "But perhaps I have been too quick, too eager, to give them the answer. This is 8th Grade stuff, now I shall show you".

The frame factors which obstruct all mathematics education:

"... this dilemma about coming through the book and to stop and think" ... "They need to train the operations, so as to be properly fixed".

In the end:

"I have started thinking, does one kill something in this way?"

This text should provide some background about Ivar. He now has to cope with the tasks of the course, and is interviewed again towards the end of it. He starts by saying that he has been influenced in a positive way.

"Positive because I have become convinced, both by what I have heard and by the small experiments I have done. And I was cautious about being influenced".

More specifically:

"Let the pupils try first, and come in as a teacher later, rather than the other way round".

This has resulted in:

"In any case I have used more time on 'the pupil side'".

What about the future? Will Ivar adopt some of the new principles promoted at the course in some systematic way?

"We shall see. I cannot work so systematically with them. Spend so much time. I do not know if I can be systematic in working with the subject matter in everyday teaching".

As a project task he chose to analyze his 8-year-old pupils' methods of addition. He notices that, contrary to textbook tradition, the pupils: "Take the tens before the ones".

He decides to follow up this strange result by introducing a subtraction problem to four pupils. He joins the pupils using a tape recorder (a nice side effect of the first interview).

"I had to help them, direct them in a way, and then I do not know how interesting the results are".

"When I listened to the tape, I noticed that I may have given them too many hints. How much of the solution was mine. Yes".

Still he notices the pupil who subtracts the "wrong" way, from left to right.

"His logic is proper. ... But then I thought about the future, perhaps this is not good, what about his future teachers?".

The pupil's algorithm causes also another sort of dilemma with Ivar:

"I believe it is more difficult to explain his method. He understands it. ... But it is more simple to me to explain the other method".

This latter statement of Ivar is one of many incidents which contribute to the interpretation that Ivar's desire to be some sort of performing master in front of the class acts as a major control of his didactical thinking. He expresses reluctance when concluding his written thesis:

"This project has provided me with useful insight about alternative ways of thinking about mathematics education — especially about subtraction. I feel, however, it is difficult to tell how this might be used when some subject matter has to be presented in plenary in class".

As an Evaluation of the Course.

My major interest with these interviews was to gain some insight about teachers' didactical concept of mathematics knowledge, in order to assess the form and the content of our courses in teacher education. One can argue here that an analysis such as this is not an evaluation of the teacher education which Ivar was offered. One could argue that Ivar's psychological structure, his Ego if one wants to use that word, blocks his desire to release both algorithms and their uses to his pupils. His desire to be the performer in the classroom, the one who demonstrates and explains, might make it impossible for him to let his pupils take control of knowledge.

Interpreted this way his reluctance is reduced to an individual psychological problem. The teacher educator can do nothing besides changing the course completely, in order to match Ivar's personality.

On the other side, the apparent resistance may be interpreted as a result of insecurity and anxiety. If this is so, the observed reluctance is a proper expression of insecurity, rather than a show of some opportunistic play with the teacher educator about something the interviewee heartily disagrees with. If such an insecurity is present, it shows potential for development if treated properly. This would imply that it may be worth transforming the course to match Ivar's didactical thinking.

The course would then need to be changed both at form level (i.e., general approach) and at content level.

Form: The course could start with less radical, more open discussions, providing for small experimental designs in the classroom, rather than guiding the student teachers into larger project works, etc.

Content: The classes could analyze and discuss issues which Ivar brings into the interview: "Explanation", "logic", "development of thinking", "lack of time".

4. SOME GENERAL PRINCIPLES OF PRACTICAL HERMENEUTICS

Practical hermeneutics is an interpretational activity. It is a method which, according to Dilthey and Schleiermacher, aims at understanding an individual, an author, better than he understands himself. It is a method which, according to Stigliano (1989), aims at challenging what is taken for granted, by turning it into a question. To the mathematician it is an intriguing method of assessment, because it is based on the interpretation of texts, that is qualitative data, rather than on quantitative data (Gadamar, 1988, Ricoeur, 1980–85).

How are such texts produced? Mainly by conversations with a student which are taped. The text is produced by means of the tape. As practical hermeneutics is used as a research method more extensively now in education, videotapes are also used for this purpose to overcome some of the limitations of written text.

I have deliberately used the word "conversation". "Discourse" is another word. We may have used the notions "qualitative research interview" or even better "interactive interview" to show the joint participation between the interviewer and the interviewee during the interview.

By careful interpretation of such texts it is possible to get insight about the understanding of an individual about some phenomenon. The advantages of such a method as compared with experimental designs within education, psychiatry and psychology, have prepared the ground for a new research tradition, *Human Science Research* (HSR)[5].

It is possible to interpret a text according to two strategies.

1. The researcher, or the interviewer, establishes his own theoretical platform, and develops the conversation according to it. The interview then functions as a tool of assessment of the platform. Is it possible to communicate with the interviewee according to the relations which constitute the platform? Does, for instance, the mathematics educator have a theoretical framework and is it presented in a form that makes communication with the student teacher possible?

In my research project I have tried to interpret the texts from the perspectives of radical constructivism and goal-directed activity theory. Did I manage to develop any meaningful conversation by implementing relations derived from these didactical platforms? Did the interviewees avoid such conversation by bringing in other issues? As long as I structure the conversation according to my intentions and my motives, I give the interviewee the status of object. What the interviewees says is interpreted through the glasses I wear.

2. The researcher, or the interviewer, makes every possible attempt to step back during conversation and interpretation in order to reduce the effects of his presuppositions. An obvious way to do this during the interview is to let the interviewee direct the conversation, that is, bring in the issues she wants to talk about. When interpreting the texts it is important to read them in an unbiased way. It is the texts that shall produce meaning, not the interpreter's expectations to them[6]. Many researchers think that several readings under shifting circumstances promote such an unbiased interpretation. Reading the same text at a beach, in a train, and in a café helps one to look at the text with fresh eyes, and to discover what might be in it. Such a reading practice thus helps to promote the interviewee as a subject.

This is in harmony with the HSR-tradition. This tradition claims that its goal it to obtain insight about the understanding of people of some phenomenon that is experienced in their world. In order to achieve this, it is necessary for the researcher to contain his own conceptual presuppositions.

I have tried to read the texts I have produced together with my teacher students in both ways. I am not sure, however, that it is possible to interpret the same text according to both positions. The production of the text already includes interpretations by interviewer and interviewee. So what stance did the interviewer take during the interview, did he try to get feedback about his own ideas, or did he mainly follow up what the interviewee wanted to say?

In my case I tried to do both.

I have interviewed 20 experienced teachers and analyzed the texts in both ways. I will give an example of a general result that includes an assessment of the course from which the texts were taken. The texts reveal that the teachers experience difficulties in developing their teaching when they see different teaching methods that contradict, rather than support each other. In the teachers' understanding of mathematics education there are two distinct ways to handle mathematical knowledge:

Old way — New way
Traditional way — Alternative way
Everyday mathematics — Experimental mathematics
Textbook work — Work freely

The "old" way is work with textbooks, mainly solving textbook problems in order to obtain adequate skills with respect to examination. The "new" way is what is recommended by the sort of teacher education which in various ways transcends the problem solving in the textbook. The "old" way as described by the teacher has also been described by mathematics educators. This tradition is based on problem solving. The tasks (exercises) to be handled by the students can usually be characterized in the following way: 1) they are ordered linearly, and usually labeled by a natural number, and 2) each task is closed such that as the student solves it he is guided to the next task (Christiansen, et al., 1984).

The striking issue here, which has not been taken with sufficient seriousness by those designing the courses, including myself, is the neglect of handling different educational practices in their didactical totality. How can, for instance, project work in mathematics support work with routine tasks, and how can work with routine tasks support project work?

Rather than demonstrating to student teachers all the nice "new" ways, much experience points to the importance of helping teachers to relate "new" ways to "old" ways in a manner that also includes practice according to "old" ways. In the texts, nothing shows teachers how to coordinate the "old" and "new" into a totality. This must mean that the teacher education they are offered does not help them in any effective way to develop their own teaching strategies, as these are related to "old" ways. Thus, the teachers are asked to learn what the education promotes, then to get fresh ground under their feet, and start practising in a completely new way.

After having worked with such problems for more than a decade, it strikes me that mathematics education is weak when it comes to handling a multitude of methods and perspectives in such a way that they are organized as parts of a totality by teachers and students. In which ways can the "old" contribute to the "new" and conversely? In the teachers' texts there exists an antagonistic contradiction between the "new" and the "old". We still do not know very much about the possibilities of relating "old" and "new" to each other such that each can function as a strength for the other.

5. SUMMARY ON METHOD

The problem of validity with practical hermeneutics has been discussed widely. If this method is used as a tool for assessment in order to give the student the status of subject, the first question which arises is: How can we know that our interpretations are valid or how can we be confident that the implications we draw from them will be operational when we revise curricula or teaching?

Kvale (1989) points to the criticism of qualitative research in general.

"Qualitative studies have been rejected as subjective, unreliable and invalid. In particular, judged against the common criteria of validity, as developed for psychometric tests, the qualitative methods have been dismissed as unscientific".

Kvale points out that there is no validity in the interview *method* as such, it is the *results* of an interview study which must be validated in a concrete situation.

Leahay, in his well written and clear review about the methodology of the HSR-tradition, stresses the principle of cooperation between interviewer and interviewee (Leahay, 1989). This cooperation should start during the interview as the interviewer and interviewee help each other to a common understanding of what the interviewee says. He quotes Kvale:

"... the interviewer during the interview condenses and interprets the meaning of what the interviewee describes, and may 'send' the interpreted meaning back".

This return of interpretations is repeated until there is, ideally, only one interpretation left.

The interviewer will certainly find new meanings later when he reads the transcript of the interview. By then he can assess the depth of this meaning by examining how it relates to the text as a whole. That is, interpretations of some part of the text have to be done in context of the text as a totality. The interviewer also has the possibility of a second (and third, and fourth ...) interview by which he can have his interpretations confirmed.

In general there exist two stages of interpretation. *Stage 1* is interpretation of the text together with its author, and *Stage 2* is interpretation of the text without the author (interviewee). The interpretation presented in my study was made without an author.

The validity of interpretation during Stage 2 is based on communication and cooperation between the interviewer and the interviewee. The validity of Stage 2 interpretation is more difficult to maintain. It is based on the interpreter's *reconstruction* of the meaning of the text. Within the HSR-tradition the validity of this interpretation will be categorized as *communicative validity*, or the interpretation's strength in relation to alternative interpretations. For discussions about such validity (which is deeply rooted in modern hermeneutics) see Kvale (1989).

Leahay (1989) has said that,

> "from what has been presented so far we can responsibly think that the knowledge we get from research interviewing is *co-constituted in the interaction between interviewer and interviewee*".

As a means of assessment which shall function as a guideline for course development and qualitative changes of teaching and learning, this principle becomes self-evident. Considering the student as a subject, it is hard to see how the student's relation to learning can be developed if not through dialogue and cooperation about assessment.

NOTE

1. I Include Vygotsky as a founder of goal-directed activity theory. His book (1978) contains a consistent, although indirect, use of the notion of "activity". See also Wertsch (1981) and Mellin-Olsen (1987).

2. Freire's work can be interpreted as a strategy to overcome political resistance and avoidance in education by politicizing teaching. His articles, 1970a and 1970b, are clear examples of this.

3. It may be profitable to establish Diderot as an ancestor of didactics. He was a chief editor of the *Encyclopédie* and thus a director of the largest project of communication of knowledge during the Enlightenment. In addition to this he also wrote several essays about the relationship between man and knowledge, which among other things, contain the ideas of modern constructivism (cf. Bremner 1983, Diderot 1967).

4. During the late 1970s and the 1980s, sociologists developed a new way of thinking about the reproduction of society. Paul Willis, who did field work among boys from English working classes, was struck by how active they were in the reproduction of their culture. He found it impossible to stand by the reproductionist theories of the 1960s and 1970s, and worked out a theory of production of culture. He was followed by several others, such as Henri Giroux (1983) and Anthony Giddens (1979).

5. A group of psychologists and psychiatrists have developed the main HSR-tradition. Annual research conferences are now arranged. Main publications include Kvale (1989), Mischler (1986) and Polkinghome (1988).

6. Amadeo Giorgi is the one who has worked in depth on this issue. He criticizes the mainstream tradition of psychological research for framing research results through design. Although it is not possible to fully eliminate this influence, the researcher should at least take into account how he constitutes the research situation, and try to control this. In order to break with this mainstream tradition, Giorgi has developed a phenomenological psychology (Giorgi 1970, 1985).

REFERENCES

Bremner, G.: 1983, *Order and chance, the pattern of Diderot's thought*, Cambridge University Press, Cambridge.

Brousseau, G.: 1987, *Theorétisation des Phénomènes d'Enseignement des Mathématiques*, Thèse, Université de Bordeaux.

Christiansen, B., Howson, A.G. & Otte, M. (Eds.).: 1984, *Perspectives on mathematics education*, Reidel, Dordrecht.

Diderot, D.: 1967, *The Enclyclopedia*, Harper and Row, New York.

Freire, P.: 1970a, 'The adult literacy process as cultural action for freedom', *Harvard Educational Review* **40**.

Freire, P.: 1970b, 'Cultural action and conscientization', *Harvard Educational Review* **40**.

Gadamar, H.G.: 1988, *Truth and method*, Sheed and Ward, London.

Giddens, A.: 1977, *Studies in social and political theory*, Hutchinson, London.

Giddens, A.: 1979, *Central problems in social theory: Action, structure and contradiction in social analysis*, University of California Press, Berkeley, CA.

Giorgi, A.: 1970, *Psychology as a human science*, Harper and Row, New York.

Giorgi, A. (Ed.): 1985, *Phenomenology and psychological research*, Duquesne University Press, Pittsburgh.

Giroux, H.: 1983, *Theory and resistance in education*, Heineman, London.

Kvale, S. (Ed.): 1989, *Issues of validity in qualitative research*, Studentlitteratur, Lund.

Leahay, J.: 1989, *The co-constitution of knowledge issued from qualitative research interviews*, HSR-Conference, Department of Psychology, University of Aarhus (Denmark).

Lenné, H.: 1969, *Analyse der Mathematikdidaktik in Deutschland*, Klett, Stuttgart.

Mellin-Olsen, S.: 1981, 'Instrumentalism as an educational concept', *Educational Studies in Mathematics* **12** (3), 356–67.

Mellin-Olsen, S.: 1987, *The politics of mathematics education*, Reidel, Dordrecht.

Mischler, E.G.: 1986, *Research Interviewing*, Harvard University Press, Cambridge, MA.

Polkinghome, D.E.: 1988, *Narrative knowing and the human sciences*, State University of New York Press, Albany, NY.

Ricoeur, P.: 1980, *Hermeneutics and the human sciences. Essays on language, action and interpretation*, Cambridge University Press, England.

Ricoeur, P.: 1980–85, *Temps et récit I-III*, Paris.
Stigliano, A.: 1989, 'Hermeneutical Practice', *Saybrook Review* **17**.
Vygotsky, L.S.: 1978, *Mind in society*, Harvard University Press, Cambridge, MA.
Wertsch, J. (Ed.): 1981, *The problem of activity in Soviet psychology*, M.E. Sharpe, New York.
Willis, P.: 1977, *Learning to labour*, Saxon House, Farnborough.
Willis, P.: 1981, 'Cultural production is different from cultural reproduction is different from social reproduction is different from social production', *Interchange* **12**.

Stieg Mellin-Olsen
University of Bergen,
(formerly at Bergen College of Education),
Norway

HERBERT P. GINSBURG & SUSAN F. JACOBS & LUZ S. LOPEZ

ASSESSING MATHEMATICAL THINKING AND LEARNING POTENTIAL IN PRIMARY GRADE CHILDREN

1. INTRODUCTION

In this paper we make three arguments concerning assessment in mathematics education: Teachers need to assess students' thinking and learning potential, not just the rote memorization of number facts or ability to use calculational procedures in a mechanical fashion. Researchers have already developed assessment methods of this type and they can be adapted for classroom use. The resulting assessment of students' thinking transforms classroom practice, leading to the "thinking curriculum". Although our work has focused on the primary grades, we believe that our arguments are relevant for children at all levels of schooling.

2. THE NEED FOR NEW APPROACHES TO ASSESSMENT

In response to widely recognized social concerns and stimulated by new research in cognitive psychology, teachers and researchers are beginning to recognize both the need and the potential for significant change in the teaching and assessment of school mathematics.

Teaching students to get the right answer in the shortest possible time with the least possible amount of thinking is no longer a useful goal in mathematics. Calculators and computers can execute tedious calculations far more efficiently than can humans.

Instruction and assessment are inseparable. If our teaching stresses rote learning, then our tests must attempt to measure it. And if we test for rote learning, using the results to evaluate students, teachers, schools, or nations in a competitive setting, then our teachers inevitably prepare their students for the test. Teaching to the test is obviously most destructive when the test measures trivialities unrelated or antithetical to desired curriculum goals. But to the extent that assessment attempts to measure interesting aspects of thinking and learning, it can facilitate instruction (Frederickson & Collins, 1989).

At the present time, assessment is still dominated by standard tests focusing on rote learning and mechanical use of procedures — rather

M. Niss (ed.), Investigations into Assessment in Mathematics Education, 157–167.

uninteresting products of the student's learning. They are inadequate to the task of revealing what the teacher most needs to understand in order to plan instruction: the student's thinking processes and strategies, and the student's learning potential. Fortunately, cognitive psychology has provided insights into the measurement of thinking and learning pointing the way to new approaches to classroom assessment.

Several alternatives to norm-referenced standardized testing include a range of interview methods, self-report procedures, and techniques for assessing learning potential. These thinking-oriented approaches to assessment allow the teacher, at all grade levels, to create a rich and practical theory of the individual student. Moreover, their use fosters a classroom climate hospitable to the implementation of a "thinking curriculum" in which both teaching and assessment contribute to intellectual development. Although these methods can be used in connection with any of the academic subjects, our focus is on children's mathematical thinking.

3. WHAT DO WE WANT TO ASSESS?

One of Piaget's most important legacies is a focus on children's thinking. He taught us to examine the processes of children's mental activity, not its products. Piaget himself came to this insight as a young man when he was engaged in administering a standardized IQ test to young children. He soon became fascinated with the reasons for children's answers, right or wrong. He was less interested in the answers themselves than in the sometimes marvelous and apparently strange processes which produced them. Just as Freud, Piaget's early inspiration, attempted to dig below the surface of the manifest dream to uncover the cognitive "dream-work" beneath it, so Piaget attempted to discover the hidden thinking responsible for the IQ test answers (Ginsburg & Opper, 1988, p. 3). Piaget's investigations led to the conclusions that children's thinking is often different from the adult's and that even a "wrong" answer may result from interesting — and indeed "logical" and sensible — thought processes.

For our purposes, details of Piaget's findings, which after all do not concern thinking about academic subject matter, are less important than his general goal of uncovering and describing basic thought processes. Adopting this goal, we must attempt to gain insight into the thinking that produces the student's everyday mathematical behavior in the classroom. In doing this, we have to move beyond the particular tasks Piaget devised; our interest is in determining how the student accomplishes column addition or the equivalence of fractions, not in investigating the cognitive processes responsible for performance on Piaget's classic conservation or classification tasks.

4. HOW CAN WE ASSESS MENTAL PROCESSES?

How can we gain insight into student's mental processes? It is unlikely that we can learn about these interesting features of understanding simply by recording and evaluating the correctness of a student's answers to standard test problems like 3 - 2. Consider other types of assessment that may be useful: the clinical interview method and its derivatives.

The Clinical Interview and Its Derivatives

The essence of standardized testing is the constraint that the interviewer administer the test to all children in as constant and unvarying a manner as possible. This means asking all children the same questions using the same words. Under conditions of standard administration, the tester is not allowed to rephrase the question if the child does not seem to understand it, and the tester is certainly not allowed to pose different questions to different children. After all, it would be unfair to treat some children differently from others.

In the course of his work with IQ tests, Piaget concluded that standardized testing — despite its laudable goal of treating all children fairly — often provides little insight into children's thinking. The main weakness of standardized testing, according to Piaget, is its very essence (and in a sense its strength) — its requirement that all children be treated in the same manner. If this be done, then the examiner cannot effectively deal with misunderstanding of questions and cannot probe for methods of solution.

Given these weaknesses inherent in standard testing, Piaget (1929) developed the *clinical interview* method, which we term "flexible interviewing", and which we believe is the most informative — and difficult — assessment method available for the purpose of assessing mathematical thinking.

Flexible interviewing. The essence of this kind of interview — which can be used to investigate almost any kind of thinking — is its flexible, responsive, and open-ended nature. Although at the outset the interviewer has available several tasks likely to be appropriate for the topic at hand, initial questions are intentionally quite general, allowing the student to determine the direction and content of the interview. Specific tasks and questions presented to the student as the interview progresses are determined by the student's responses to the initial questions and tasks.

The interviewer is responsible for establishing rapport, preparing a series of appropriate tasks, and listening to and observing the student's responses. The interviewer's goal is open-ended. Topics of inquiry and issues under investigation may not have been planned or even anticipated prior to the interview. Because they have been suggested by the student's responses to initial questions, the issues investigated are likely to be of immediate interest or concern to the student. The observant and thoughtful interviewer constantly adapts interviewing strategy to the direction suggested by the student.

Non-specific questions such as "How did you do it?" or "What did you say to yourself?" or "How would you explain it to a friend?" encourage rich verbalization but give no suggestion of how the student should respond. Students should be encouraged to work with chips or tiles, pencils or marker pens and paper, and small toys. These manipulative objects motivate the student and because they tend to externalize thinking allow the interviewer-observer additional insight into thinking processes.

Responding to the student's responses and behavior, the interviewer makes and tests hypotheses about the student's thinking. Tasks are varied and modified, becoming more specific in order to focus on particular aspects of thinking, and more difficult in order to test the limits of the student's understanding.

Flexible interviewing is non-standardized in many ways. Establishing rapport requires warm encouragement and calming of apprehension for some students. For others, a tone of high expectation produces better results. While one student can be allowed to follow a train of thought at length, another student will need closer control, with more frequent questions. Questions must be phrased so that the student understands them. A skilled interviewer is aware of the student's personal vocabulary and can rephrase questions in the student's words.

The interviewer is not satisfied with a "correct" answer, but tests the strength and consistency of the student's beliefs by use of repetition and counter-suggestion. An "incorrect" answer may reveal valuable information about thinking processes. Whether "correct" or "incorrect", the most desirable answer is one which provides the most information concerning the student's thinking.

Obviously, maintaining the student's interest is necessary if the student's best, rather than ordinary performance is to be measured. The bored, distracted, tired, or uncomfortable student will not reveal much useful information. When the student is interested in the issues, enjoying the attention of the interviewer, and when the level of difficulty of the task is challenging but not overwhelming, a rewarding dialogue results. The interviewer may learn not only about the student's thinking, but also about important attitudes and beliefs concerning the student's own ability, and concerning the goals, methods, and nature of mathematics.

Self-reports are often requested during a clinical interview. The most commonly used self-report procedures are thinking aloud and retrospective reports (Ericsson & Simon, 1980). The interviewer usually finds it useful to ask the student to "think aloud" or "talk aloud" while he or she solves a problem. Through this "on-line" or concurrent verbalization valuable information about the student's thinking processes can be obtained.

Similarly, the interviewer often asks the student to verbalize his or her thinking immediately after the completion of the task rather than during its execution. The interviewer will ask a question such as, "Tell me everything you were thinking while you were solving the problem", or "Tell me how you solved the problem". The information obtained from these two procedures complements and enriches the flexible interview.

Consider one authentic example of flexible interviewing. This example shows that a correct response may lie on a foundation of unsound understanding. The interviewer observed that Chris wrote the correct answer, 14, when given 9+5.

"I: Now I'm going to ask you something about 14. How come you wrote 14 with a 1 and then a 4?
C: 'Cause that's how I write 14.
I: I notice that when you write 14 you have a 1, and on the right of that is a 4. What does that 4 stand for?
C: 'Cause it's 14.
I: All right. What does the 1 stand for?
C: That's how you write 14.
I: Why don't you write it like this [41]?
C: That's 41.
I: All right. Why do you write 41 like that?
C: Because there's a 4 there and a 1 there.
I: Why did they invent that way of doing it? What could it possibly mean? What does that 1 stand for?
C: 1.
I: What does that 4 stand for?
C: 4.
I: Can you write the number 123? That's right. That's how people write 123. What does that 1 mean?
C: 1
I: Just 1. And what does that 2 mean? What does it stand for? What is it telling us?
C: 2.
I: Just 2. And the 3? What does that tell us?
C: Just 3."

Clearly, Chris did not understand the place value meaning of the symbols he wrote accurately.

Organized probes. Another, more structured variation on clinical interviewing is the "organized probe". Ginsburg (1990) has developed a comprehensive series of organized probes in connection with the *Test of Early Mathematics Ability* or *TEMA* (Ginsburg & Baroody, 1990), which focuses on various aspects of informal and formal mathematical knowledge in children from preschool through approximately Grade 3. The test items were selected from data gathering procedures used in cognitive development research on mathematical thinking over the past 20 or 30 years. The test items deal with such matters as informal addition; the mental number line; the concept of cardinality; simple number combinations; alignment procedures for addition and subtraction; and base ten concepts.

Although the TEMA covers interesting areas of mathematics performance, it does not shed a great deal of light upon thinking processes generally and understanding in particular. Consequently, Ginsburg (1990) developed an organized system of probes to be used in parallel with the TEMA. The general idea was that after the TEMA had been given in the standard fashion, many examiners would find it useful to probe further into the thought processes which produced the observed performance, particularly in the case of errors. Most evaluators, however, have not had training or experience in assessing students' thinking. Consequently, Ginsburg attempted to provide examiners with a structured and comfortable procedure for probing the strategies and concepts underlying students' responses to the TEMA. Of course, clinical interviewing is a more effective and difficult means for achieving the same purpose; yet because most evaluators are not prepared to engage in extensive clinical interviewing, organized probes are a useful first step.

The probes for each of the 65 items of the TEMA first attempt to establish whether the student has understood the basic question. Often students produce an incorrect response because they have misinterpreted a minor feature of the question asked. The probes attempt to distinguish this situation from that in which students do not understand the question because they fail to comprehend the relevant concept.

For example, one of the TEMA items attempts to deal with the addition of multiples of ten by asking, "Here are some questions about adding money. We'll pretend that you have some money and I give you some more. If you start with nine dollars and I give you one ten-dollar bill, what do you end up with?" The wording was of course intended to be clear in the first place, but some children misunderstand the question, perhaps suffering from confusion about the phrase, "What do you end up with?" In a true clinical interview, the examiner would attempt to determine what each child fails to understand and then re-word the question accordingly. In the system of probes, on the other hand, the examiner has available two alternative questions. One attempts to deal with the "What do you end up with?" problem by substituting the following: "If you have nine dollars and I give you one ten-dollar bill, how much do you have altogether?" A second question is intended for children who may be confused about money: "If you start with a nine, and then you add on a ten, how much do you end up with?" Given such alternative questions some children suddenly see the light, exclaiming, "Oh, is that what you meant!"

Next, the probes attempt to determine the strategies and processes used by the student to solve the problem. For example, in the case of concrete addition, the basic TEMA question is: "Joey has two pennies. He gets one more penny. How many does he have altogether? If you want, you can use your fingers to help you find the answer". As the questions are asked, the examiner shows the child the numbers of pennies involved. The probes then attempt to determine whether the student used such procedures as counting on the fingers, mental counting on, or memorized number facts. Three types of questions are suggested. The first is "It's OK to use your fingers. Put your hands on the table and show me how you do this. Tell me out loud what you are doing". In this case, some children who had previously thought that it is improper or even cheating to use the fingers breathe a sigh of relief and show the examiner their finger counting procedure. A second question is: "How did you know the answer was ___?" In this case, some children say, "I know that three and two is five because I just knew the answer. I learned it". If the answer was given quickly, this may be a good indication of the use of memorized facts. A third question is: "Why does six pennies and two make ___ altogether?" In response to this question, some children say, "It had to be eight because I know that six and one is seven and then one more must be eight". Here the use of a reasoning strategy is evident.

5. TOWARD THE THINKING CURRICULUM

Clearly, the various assessment methods can be used in classrooms to obtain important information concerning students' thinking. But using them also affects students in important ways and transforms the way in which teaching is conducted.

How Thinking Assessment Affects Students

The focus on thinking can provide students with several valuable lessons. One defines the "curriculum".

Thinking is taught here. The teacher's focus on the student's thinking says in effect: "I value your thinking, your way of dealing with problems. I am interested in your thinking; it is the center of my attention. Your thinking is so important that I must understand it. Your thinking is so powerful that it determines how I teach; it influences what happens in the classroom. In this class, we do not just memorize material; we solve problems by thinking. We always use our minds and we also think about using our minds. Thinking is taught here".

Meta-cognition. Another lesson involves training in meta-cognition. As an example, consider the retrospective procedure, described earlier. Recall that in a 2nd Grade class discussion centered on students' descriptions of their methods for solving simple computational problems. Throughout this process, the teacher established an atmosphere in which students were encouraged to share their strategies, to value the range of strategies arising in the classroom, and to verbalize strategies as explicitly as possible. The teacher showed the class that there can be a variety of ways to obtain the correct answer. After exposing the students to a number of strategies, she asked them if they would like to choose a different strategy and if so why they would choose it. Some students preferred to continue in the strategy already chosen, while other students chose a new approach, often "because it is faster". In any event, although the assessment procedure provided the teacher with valuable information concerning students' thinking, it also provided the students with a lesson in what thinking is all about. In effect, the lesson involved the sharing, valuing, and teaching of thinking about thinking.

The assessment methods are interactive, requiring verbal communication. The thinking curriculum helps the students to express their thought, to put it in words. Expressing thought requires students to focus attention and reflect upon their own problem solving processes. The assessment methods help students to develop meta-cognitive skills used to monitor, check, and evaluate their performance. The general lesson is: "meta-cognition is taught here".

Some Effects on Teachers

Recently one of the authors had the opportunity to train a group of teachers in the country of Colombia on how to use some of the thinking assessment techniques. Teachers were trained to use thinking-aloud procedures, retrospective reports, and structured clinical interviewing techniques to assess students' thinking. These techniques gave a new dimension to the teachers' work. The teachers used these assessment techniques in the classroom not only to investigate thinking, but as tools to foster it.

The Teacher as Thinker

The thinking curriculum requires the teacher to think, too. To teach thinking, the teacher has to model thinking: introspection, self-monitoring and checking, and weighing of alternative strategies. At the elementary school level, the experienced mathematics teacher probably knows the material so well that thinking aloud, verbalizing weighing of strategies, and introspection are in a sense staged for the students in order to help them learn these meta-cognitive skills for themselves. When there is a real mathematical question in the teacher's mind, the process acquires greater urgency and greater potential for learning. Challenging questions may inspire curiosity and interest in follow-up work. Sometimes a student will offer a step towards solution of the problem. If the teacher respects this offer, even if it is imperfect or unclear, the students can learn to see the value of exploratory thought. As a thinking model, the teacher may admit uncertainty and ask the student to rephrase the question or in other ways try to elicit clarification. In this small but real world situation the students observe the methods their teacher uses to grapple with a problem.

It is important for students to know that adults do not have ready-made answers on all occasions. Many times during the day teachers have to make decisions and plans that affect their students. The teacher's weighing alternatives and thinking aloud gives students indirect experience in the decision-making or planning process. As students are encouraged to enter into this process with their own comments and suggestions, the learning experience becomes more direct. When students actually provide helpful, imaginative, and constructive ideas, the teacher and students benefit together.

The Teacher's Sense of Self As A Teacher

In Colombia, the assessment techniques fostered reflection in the teachers. Prior to our joint work, they viewed themselves as authority figures. Along with this view was the belief that they should present themselves to students as people who know everything and do not make mistakes. As our work developed, teachers began to see the importance of presenting themselves as human beings to their students. Teachers even began to model confusion and making mistakes in the classroom.

This implied that the teachers' sense of the nature of mathematics education had changed. They down-played their emphasis on right or wrong answers. Instead, they asked students to spell out the processes used to solve problems, and gave more credit for this aspect of problem solving than to the answer itself. The curriculum had shifted to mathematical thinking.

6. A CONCLUSION OF RECIPROCAL IMPLICATION

Classroom assessment should focus on mathematical thinking. It is now possible to do this with some success. And the assessment of thinking leads to a distinctive approach to mathematics education (and education in general). This type of assessment is based on the assumption that the core of education is the fostering of thinking. If the focus is on thinking, then the subject of the mathematics curriculum is not mathematics as it is sometimes conceived, namely a collection of correct procedures and answers which must be learned as such. Instead, the curriculum is thinking about mathematics. Mathematics education should involve helping students to think mathematically and to think about their mathematical thinking. This thinking curriculum requires a constant focus on (and assessment of) thinking, not on mathematics as a finished, static product. Thinking assessment requires the thinking curriculum and the thinking curriculum requires thinking assessment.

REFERENCES

Artzt, A.F. & Armour-Thomas, E.: 1990, 'Development of a cognitive-metacognitive framework for protocol analysis of mathematical problem solving in small groups', paper presented at the Annual Meeting of the American Educational Research Association, Boston, MA.

Budoff, M.: 1987, 'Measures for assessing learning potential', in Lidz, C.S. (Ed.), *Dynamic assessment: Foundations and fundamentals*, Guilford Press, New York.

Campione, J.C., Brown, A.L., Ferrara, R.A. & Bryant, N.R.: 1984, 'The zone of proximal development: Implications for individual differences and learning', in Rogoff, B. & Wertsch, J.V. (Eds.), *Children's learning in the zone of proximal development. New Directions for Child Development*, Jossey Bass, San Francisco.

Carlson, J.S. & Widaman, K.F.: 1986, 'Eysenck on intelligence: A critical perspective', in Modgil, S. & Modgil, C. (Eds.), *Hans Eysenck: Consensus and controversy*, The Falmer Press, Philadelphia.

Cocking, R.R. & Chipman, S.: 1988, 'Conceptual issues related to mathematics achievement of language minority children', in Cocking, R. & Mestre, J.P. (Eds.), *Linguistic and cultural influences on learning mathematics*, Lawrence Erlbaum, NJ.

Dionne, J.J. & Fitzback-Labrecque, M.: 1989, 'The use of "mini-interviews" by "orthopédagogues": A three case studies', *Proceedings of the eleventh annual meeting of PME-NA*, New Brunswick, NJ.

Ericsson, K.A. & Simon, H.A.: 1980, 'Verbal reports as data', *Psychological Review*, **87**(3), 215–151.

Feuerstein, R.: 1979, *The dynamic assessment of retarded performers: The learning potential assessment device, theory, instruments, and techniques*, University Park Press, Baltimore.

Frederickson, J.R. & Collins, A.: 1989, 'A systems approach to educational testing', *Educational Researcher*, **18**, 27–32.

Ginsburg, H.P.: 1990, *The test of early mathematics ability: Assessment probes and instructional activities*, Pro-Ed, Austin, TX.

Ginsburg, H.P. & Baroody, A.J.: 1990, *The test of early mathematics ability* (2nd ed.), Pro-Ed, Austin, TX.

Ginsburg, H.P. & Mathews, S.C.: 1984, *Diagnostic test of arithmetic strategies (DTAS)*, Pro-Ed, Austin, TX.

Ginsburg, H.P. & Opper, S.: 1988, *Piaget's theory of intellectual development (3rd ed.)*, Prentice-Hall, Englewood Cliffs, NJ.

Johnston, W.B. & Packer, A.E. (Eds.): 1987, *Workforce 2,000: Work and workers for the twenty-first century*, Hudson Institute, Indianapolis, IN.

National Council of Teachers of Mathematics: 1989, *Curriculum and evaluation standards for school mathematics*, Reston, VA.

National Research Council: 1989, *Everybody counts: A report to the nation on the future of mathematics education*, National Academy Press, Washington, DC.

Newell, A. & Simon, H.: 1972, *Human problem solving*, Prentice-Hall, Englewood Cliffs, NJ.

Peck, D.M., Jencks, S.M. & Connell, M.L.: 1989, 'Improving instruction through brief interviews', *Arithmetic Teacher*, **37**, 15–17.

Piaget, J.: 1929, *The child's conception of the world*, Harcourt, Brace & World, Inc., NY.

Polya, G.: 1945, *How to solve it*, Doubleday, Garden City, NY.

Schoenfeld, A.H.: 1983, 'Episodes and executive decisions in mathematical problem solving', in Lesh, R. & Landau, M. (Eds.), *Acquisition of mathematics concepts and processes*, Academic Press, New York.

Schoenfeld, A.H.: 1985, *Mathematical problem solving*, Academic Press, Orlando, Fl.

Silver Burdett & Ginn: 1991, *Mathematics: Exploring your world. Alternative assessment activities*, Silver Burdett & Ginn, Morristown, NJ.

Stigler, J.W.: 1988, 'Research into practice: The use of verbal explanation in Japanese and American classrooms', *Arithmetic Teacher*, **36**, (2), 27–29.

Swanson, H.L.: 1991, 'Issues and concerns in the assessment of learning disabilities', in Swanson, H.L. (Ed.), *Handbook on the assessment of learning disabilities*, Pro-Ed, Austin, TX.

Vygotsky, L.S.: 1978, *Mind in society: The development of higher psychological processes*, edited by Cole, M., John-Steiner, V., Scribner, S., and Souberman, E. Harvard University Press, Cambridge.

Herbert P. Ginsburg & Susan F. Jacobs & Luz S. Lopez
Teachers College, Columbia University
New York
USA

BENGT JOHANSSON

DIAGNOSTIC ASSESSMENT IN ARITHMETIC

1. INTRODUCTION AND BACKGROUND

This article describes the results of different types of assessment in mathematics education. It shows how a diagnostic instrument for arithmetic has come to be used for assessing classroom instruction and for assessing national goals in Sweden. It also demonstrates how diagnostic assessment can obtain greater detail by combining the results from written tests with the results of interviews. The diagnostic assessment of students' complex knowledge and skills lags behind other areas in mathematics education.

Assessing students' knowledge and skills in arithmetic has a long tradition in Sweden. The first significant efforts came at the beginning of the 1930s from Frits Wigforss at the Kalmar School of Education. He developed two diagnostic tests, one in arithmetic and another in problem solving. He based the tests on analyses of error types and on the basic knowledge and skills necessary for students to learn to solve different problems (Wigforss, 1931, 1934, 1937). In the past 20 years, several attempts have been made in our country to develop knowledge about the diagnostic assessment of mathematics. This has taken place within the framework of research and development projects as well as in connection with a number of larger efforts in the further education of teachers.

2. RELATIVE ASSESSMENT

It is no exaggeration to say that in many countries schools exist in a culture of assessment dominated by quantitative comparisons between results. Assessment less frequently has been seen as a comparison between results and goals. I believe that this type of assessment is so seldom seen, in large part, because the goals are often unclear to the person who is constructing, conducting or interpreting the assessment. Furthermore, goals are more often created according to what it is hoped and wished that students will learn and less often according to what it is realistic to believe that they can learn under given conditions. When goals lie far above what it is possible to reach, the assessment will be adapted to *what can be reached*. The comparison made in this way will be within a very fragmentary and superficial *assessed curriculum*, not relative to a well constructed *intended*

M. Niss (ed.), Investigations into Assessment in Mathematics Education, 169–184.

curriculum. It is also much easier to compare one result with another result than to compare a result with a goal. Observing, for example, that Japanese children are better at arithmetic than Swedish children is easier than recognizing that Swedish children do all right for Swedish conditions or relative to the goals set in Sweden for mathematics instruction in schools.

Much less developed is formative or diagnostic assessment. Such a comment is not directed primarily toward describing a level of knowledge or toward comparing results and goals, but instead toward comparing results with different types of knowledge structures that the students must develop in order to reach the goals that have been set. Directed toward understanding and explaining, it is to give feedback to settings in which instruction takes place and to settings where local efforts on curricula are carried out. It aims at "getting behind" the goals and put the focus on the complex knowledge and skills that students have which either enable them, make it more difficult for them, or prevent them from reaching a particular goal.

3. AN ASSESSMENT INSTRUMENT FOR ARITHMETIC

In the middle of the 1970s, an assessment instrument was developed as a part of the *Processanalys av Undervisning i Matematik/Psykolingvistik* (*PUMP*) project (Johansson & Kilborn, 1974). This instrument includes the four standard algorithms with integers and describes similarities and differences between different types of problems with regard to number area and to the number and type of subcalculations. After reworking and simplification, it took on the following appearance.

The instrument consists of four matrices, one for each calculation method. The design is founded upon an analysis of the subcalculation that must be made when following a standard algorithm as well as on a number of empirical investigations of students' calculation skills, and theories on the capacity of the working memory (Kilborn, 1979). Each matrix consists of two parts. The upper part includes tasks that occur as subtasks in calculations in the respective standard algorithm. For example, square number 25 in the multiplication matrix (M25) contains problems of the type $6 \times 6 + 8 = ?$; such subtasks as the student meets in the multiplication algorithm when the addition of a number carried over leads to crossing over to a new column. The lower portion of the matrix includes corresponding algorithm problems, and their placement is steered by the tasks in the upper portion. The complexity of the tasks increases form left to right and from top to bottom through changes in the number and type of suboperations.

To be able to judge a student's calculation skills in a certain type of task with fairly good certainty, the instrument has been furnished with a subtest in each square that consists of four to six tasks. Kilborn has shown that the information gained on the students' calculation skills does not increase to

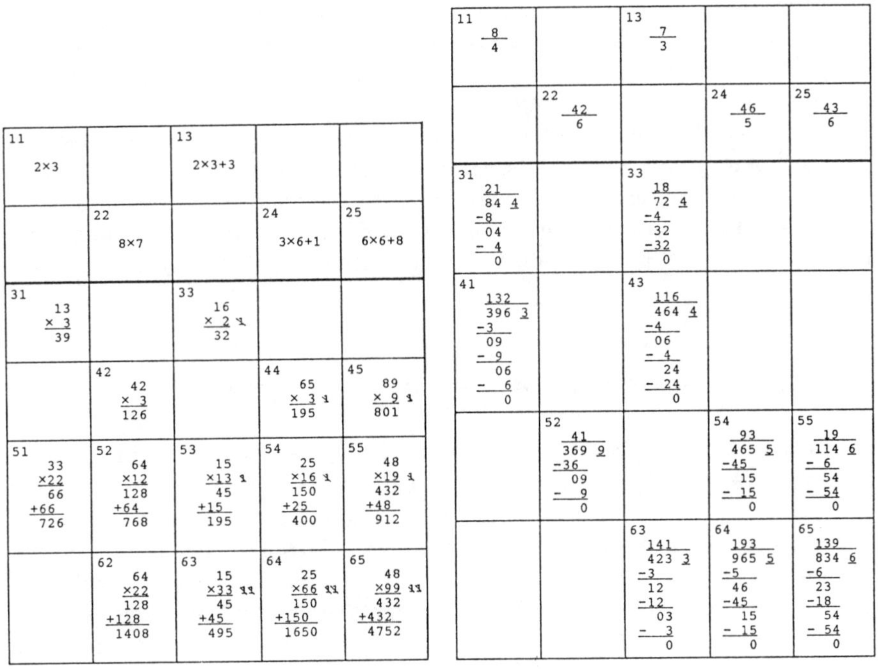

Figure 1 *Multiplication (left) and division (right) matrices*

any significant extent with a greater number of tasks, and that fewer than four tasks rapidly impairs the ability to make a statement about the students' calculation skills in problems of the type in question (Kilborn, 1979). Four to six tasks is, consequently, an attempt at balance between the requirement for reliable information and the time and effort one is prepared to give for testing the students.

In squares 11 and 22 in the respective matrices, we have collected number facts which a student meets on repeated occasions in the standard algorithm to which it belongs. For example, in a task of the type 87×96, the student must master four multiplications in a row (together with a good deal more). Thus, to be able to say anything concrete about students' knowledge of facts as a basis for their calculations in the corresponding algorithm, it is necessary to include several more tasks of the type in question than four to six. After having conducted a large number of field experiments over many years, we have discovered that a total of 15–20 randomly-selected tasks represents a good balance between the requirement for reliable information on students' knowledge of number facts, and the time one is prepared to spend on testing. A look at the number of problems solved in the tables during a limited time period, for example one minute, also yields valuable information about which students use too many lengthy ways of thinking. These students risk running into problems with their working memory when the combinations in question are included as a part of a larger task. Moreover, they run the risk of not being able to understand the relations between addition/subtraction and multiplication/-division which are based upon well-automated knowledge of the number facts.

4. ASSESSMENT OF INSTRUCTION IN ARITHMETIC

The instrument discussed here was originally developed to be used in classroom studies (Johansson, 1975; Kilborn, 1979). It primarily has been employed to judge students' basic knowledge, and skills in terms of prerequisites, and to study the design of textbooks and teaching aids. For example, with the instrument, we have been able to study a number of tasks of different kinds, where and when the tasks occur, and how the tasks have been put into sequence for different types of classroom activities. We have also been able to describe and analyze the content of the different tests in the textbooks as well as what the teacher chooses to include when teaching the whole class or different groups of students.

By taping lessons, we were able to secure nearly all the dialogues that took place between the teacher and individual students at the time of individual work with the calculations (the teacher wore a wireless microphone). In this way, we have been able to analyze, in detail, the contents of the dialogues, the students' basic knowledge and skills, and the textbooks and teaching aids that the student have worked with during these taped lessons.

A fairly depressing picture of mathematics instruction came forth in the process of this analysis, especially in the examination of all the conversations held between the teacher and an individual student while one specific unit of mathematics was taught (often stretching over five to ten lessons). Students who lacked large parts of the basic knowledge necessary very rarely received help in developing the qualities of thought needed to achieve good results. These students were, instead, often *piloted through* or prompted by the teacher past all obstacles, on the teacher's condition and according to very fixed thought patterns.

Most of the conversations were very short, rarely longer than about one minute. The larger the distance between the student's prerequisites and the demands of the textbooks, the larger the portion of the problem that was solved by the teacher became, and the more comprehensive the solution became. The time it took for the student to consider a subquestion was seldom longer than 3–4 seconds. The didactical contract which seems to regulate this dialogue (Johansson, 1975; Brousseau, 1984) forbids the student to react against instruction, even if he has not improved in calculating after the efforts of the teacher. The character of the contract was most clearly seen by the very last part of the conversation, when the teacher asked to leave the student in order to go to a classmate who also needed help. The teacher nearly always received "permission" to leave the students, in spite of the fact that the students were often left with a problem unsolved.

In interviews, it was seen that participating teachers had very little knowledge of the great number of different ways of thinking that children have and even less of how one could use different ways of thinking to reach the goals that have been set. It can be said that most teachers showed great uncertainty for work that has its starting point in the students' logic, and that they instead chose to pull the students along the path laid out by the textbook, or the teacher. In the classroom studies, the teachers did not have access to the results of the diagnoses we had made during the week before the lessons were taped.

Thus, in the situation reported here, the diagnostic instrument was used chiefly as a basis for describing and evaluating the teachers' efforts with individual students during their work with arithmetic. An example follows of the "piloting" that was so typical of the teachers' work during individual conversations with students who demonstrated serious problems with their basic knowledge. The student in this example is a 4th Grader who one week earlier at the diagnosis (which included squares 11 to 45 in the multiplication matrix) had shown that he had very poor knowledge of the multiplication facts and could not multiply a two-digit number by a one-digit number. However, like all his classmates, he was to try to learn to multiply two two-digit numbers, which was the next problem in the textbook. The conversation takes places when the teacher comes to the student's desk, 23 minutes after starting the third lesson dealing with this task. The student is to calculate 14×13 (T = teacher, S = student):

T: Let's see how you're doing.
S: It's fine.
T: Well, you haven't done very much. Multiply these numbers now.
S: Ummm.
T: How do you begin?
S: Begin with 4 times 3.
T: Hmm ... that makes ...?
S: Eight, ahh.
T: What was that?
S: Let me think.
T: Four times three ... How much is two times four?

S: Seventeen ... Thirteen.
T: How much is two times four?
S: It's eight.
T: Mm. Three times four?
S: Thirteen ... No it's fourteen.
T: No, think now. You have eight plus four ... how much does that make? Work hard now! Well, how much is it?
S: Eight plus four?
T: Mm.
S: Eight plus four makes ahhh ... nine, ten eleven, twelve.
T: Mm. What is three times four?
S: I don't know.
T: That's twelve, isn't it?
S: Oh. (Johansson, 1975, pp. 47–48)

5. ASSESSMENT RELATIVE TO NATIONAL GOALS IN ARITHMETIC

The national curriculum in mathematics in Sweden (*Lgr 80*) contains a core of tasks that all students shall have mastered before they leave the compulsory nine-year comprehensive school. The types of tasks that are included in the matrices are found in this core. As regards addition and subtraction, the students shall have mastered the greatest part of the tasks by the 3rd Grade. This is also true for limited parts of multiplication and division. The students shall have mastered nearly everything included in the matrices by the end of the 6th Grade. Thus the instrument has come to be used on different levels in our school system, from a national level to local and individual levels, in order to evaluate how well the school succeeds in its tasks of giving all students these skills and knowledge. The instrument has also come to be used as one of many associated with the question of equal educational standards in our country, an issue which has become especially important in the ongoing decentralization of the public school system.

As an example of the use of the instrument in connection with standards, some results are presented from an assessment done in a Swedish municipality. All students in the 3rd and 6th Grades (a total of approximately 7,000 students) are tested each year on a selection of tasks from the four matrices among other things. One part of the tasks occurs both in Grade 3 and Grade 6. I describe the results from some of these common tasks here. According to the goals in the national curriculum for mathematics, all students shall have mastered these tasks by Grade 3.

It can be seen from Table 1 that many students have far to go to reach the goals set forth in the national curriculum. There are very large differences between different classes, both within the same grade and between two grades. In the 3rd Grade class that shows the best results for the subtraction facts (type 13-8 = ?), for example, there is no student with

		Grade 3			Grade 6		
Type of Task	No. of tasks	Poor	Ave	Best	Poor	Ave	Best
8+7 (A22)	15	41	78	6	19	38	10
389+127 (A55)	4	26	*50*	10	13	26	4
13-8 (S22)	15	44	85	12	29	*50*	18
366-177 (S55)	4	42	67	18	24	49	7

Table 1 *Percent of students having at least two incorrect answers in the poorest, average, and best classes in Grades 3 and 6.*

more than one incorrect answer on the 15 tasks in that subtest. In the 3rd Grade class that showed the poorest results, almost 70 percent of the students had at least two incorrect answers and — which is remarkable — in the 6th Grade class that showed the poorest results, every other student had at least one incorrect answer out of the 15 tasks.

Assessments of this kind have often led to lively discussions on how results can be improved so that goals can be reached. In several cases, it has also been questioned whether the goals are actually realistic for everyone. One measure has been to allow the individuals who have responsibility for their schools to discuss probable reasons for poor results with the teachers whose classes have demonstrated such results and to investigate what support and what help, such as further teacher education, these teachers might need. In some cases, the teachers who have showed the best results have been asked to report on their successful instruction to colleagues who need to develop their professional skills. However, in the majority of the cases, little interest has been shown by the school leadership. The problem is not by any means new. It is one thing to make an assessment and quite another to deal with the consequences of the results.

As the test contains a selection of problems from the more comprehensive PUMP assessment instrument, several teachers have been able to follow up on the results themselves, and to design activities to help the students develop their calculation skills for the unit in question. This work has been made easier by the use of special student and teacher material and by an associated computer program for generating suitable tasks for students.

6. ASSESSMENT OF STUDENTS' COMPLEX KNOWLEDGE OF AND SKILLS IN ARITHMETIC

The results I have shown to this point are based primarily upon judgements of the students' performance on test, in terms of having correct or incorrect answers and in terms of the frequency of incorrect answers. In order to better understand and describe students' thinking behind all these numbers, I supplemented the assessment, in several cases, with interviews of students concerning tests and the tasks on the tests. The interviews took place both as a part of more extensive efforts to develop students' complex knowledge of and skills in mathematics, and as a part of a detailed assessment and follow-up of individual students in connection with the further education of teachers. The interviews were taped or videotaped to facilitate later analysis so that they could be used, for instance, in further education on the qualitative assessment of learning and students' thinking. As an example of this kind of individual follow-up, I present a summary of some parts of an interview with Daniel, age 14. Table 2 presents Daniel's results on the written test which he had taken a few days before, and which consisted of a selection of tasks from the four matrices:

Type of tasks	No. of tasks	No. of wrong	Type of tasks	No. of tasks	No. of wrong
8+7 (A22)	20	0	8×7 (M22)	20	10
374+837 (A65)	5	0	89×9 (M45)	6	3
13-8 (S22)	20	2	42/6 (D22)	18	11
366-177 (S55)	15	2	834/4 (D65)	14	4

Note: Each test on the different number facts (A22, S22, M22 and D22) had a time limit of two minutes. Problems for which Daniel gave no answer were classified as incorrect.

Table 2 *The number of incorrect answers by group of tasks for Daniel, 14 years.*

Swiftness and certainty in addition are sufficient for completely correct calculations in this test. In his other calculations, Daniel has problems with both number facts and their corresponding algorithms. An analysis of the types of incorrect answers indicates a way of thinking that we decided to look more closely at during an interview. The analysis showed that we should concentrate on how Daniel thinks when he tries to master the number facts in subtraction, multiplication and division. As a basis for the interviews, I used a complete description of all the facts in the respective operations, even of the parts that lie in the number area below 10 and that are found in the matrix squares S11, M11, and D11 (se Tables 3 and 4). A line in Table 4 differentiates tasks square S11 from those in square 22.

Subtraction

I began by interviewing Daniel on subtraction. The tasks were taken from the subtraction table in the order described in Table 3. Daniel answered two or three randomly chosen tasks from each group.

Task group (S11)	Example	Task group (S22)	Example
-0, -1, -2	9-1	10-	10-7
Difference 0, 1, 2	8-6	Half	18-9
Half	6-3	11-	11-5
Others under 10	9-5	-9	13-9
		Difference 9	12-3
		Others	15-8

Table 3 *The table shows the division and order of the subtraction facts that were the basis for choosing interview tasks.*

0-0	1-0	2-0	3-0	4-0	5-0	6-0	7-0	8-0	9-0
10-1	1-1	2-1	3-1	4-1	5-1	6-1	7-1	8-1	*9-1*
10-2	11-2	2-2	3-2	4-2	5-2	6-2	7-2	8-2	9-2
10-3	11-3	*12-3*	3-3	4-3	5-3	*6-3*	7-3	8-3	9-3
10-4	11-4	12-4	13-4	4-4	5-4	6-4	7-4	8-4	9-4
10-5	*11-5*	12-5	13-5	14-5	5-5	6-5	7-5	8-5	*9-5*
10-6	11-6	12-6	13-6	14-6	15-6	6-6	7-6	*8-6*	9-6
10-7	11-7	12-7	13-7	14-7	15-7	16-7	7-7	8-7	9-7
10-8	11-8	12-8	13-8	14-8	*15-8*	16-8	17-8	8-8	9-8
10-9	11-9	12-9	*13-9*	14-9	15-9	16-9	17-9	*18-9*	9-9

Note: The thicker line in the table separates square S11 from square S22 in the subtraction matrix. The tasks that were included in the interview are written in italics.

Table 4 *The table shows the division of the subtraction facts that was the basis for choosing interview tasks.*

The facts were divided on the basis of experience with previous development efforts and on recent research on students' ways of thinking in this area (Carpenter & Moser, 1984). After a short conversation concerning what he thought about his results, Daniel described his thinking when he solved the different tasks. It was difficult at first for him to express his thought, but he began to give his story more detail after a few minutes.

Daniel answered directly on the majority of tasks with numbers up to 10 without any intermediate line of thought:

"I know how to do that"

The only exception to this in the subtractions below 10 was the problem 8-5, for which he used addition:

"8-5=3 because 5+3=8"

It was especially interesting that the subtractions from the number 10 were automated. This was also true for such combinations as 18-9, 12-6 and so on:

"That's half"

On the rest of the subtractions facts he consistently used one and the same way of thinking. He translated the given problem into a two-step task that he performed with certainty, that is subtraction from 10 and addition of two units whose sum is less than 10. Thus, in this way 14-9 became 5:

"10-9 is 1, +4 is 5"

The interviewer was not the only one who understood the way of thinking to be the same in comparison with other tasks. To the problem 13-6, for example, Daniel answered:

"10-6 is 4, +3 is 7. I'm thinking the same way"

It was quite easy to understand and describe Daniel's way of thinking in subtraction. The method also explained most of the incorrect answers he had given on the subtraction test where, for example, 12-8 became 2 and 13-5 became 3. He had simply lost the results from one of the two suboperations he normally used. The judgement of the teacher who had conducted the interview was that there was no reason for trying to change the dominant way of thinking that Daniel employed. It was functional, at least in algorithm calculation and, inasmuch as he (and the teacher) had now become aware of the problem, with a well balanced training in skills, his difficulties in subtraction could soon be rectified. Daniel had not become stuck in any of the many variations of counting upwards or downwards that are often seen among weak counters (Carpenter & Moser, 1984). To try to develop Daniel's knowledge of the relation between addition and subtraction, e.g., 13-5=8, 8+5=13, would, of course, also be good.

Multiplication

As the test results showed, the students' problems were much greater in multiplication and division. The interview tasks in these cases, as in subtraction, were chosen from a group division. As the groups are identical in multiplication and division, I will only report here on the groups in multiplication. Daniel was asked to answer two or three randomly chosen tasks from each group (see Tables 5 and 6):

Task group (M11)	Example	Task group (M22)	Example
×1, 1×	1×4	×4, 4×	4×9
×10, 10×	10×3	×3, 3×	3×7
×2, 2×	9×2	×9, 9×	9×7
×5, 5×	5×6	Others	6×8
"Squares"	8×8		

Table 5 *The table shows the division and order of the multiplication facts that were the basis for choosing interview tasks.*

1×10	2×10	3×10	4×10	5×10	6×10	7×10	8×10	9×10	10×10
1×9	2×9	3×9	*4×9*	5×9	6×9	7×9	8×9	9×9	10×9
1×8	2×8	3×8	4×8	5×8	*6×8*	7×8	*8×8*	9×8	10×8
1×7	2×7	*3×7*	4×7	5×7	6×7	7×7	8×7	*9×7*	10×7
1×6	2×6	3×6	4×6	*5×6*	6×6	7×6	8×6	9×6	10×6
1×5	2×5	3×5	4×5	5×5	6×5	7×5	8×5	9×5	10×5
1×4	2×4	3×4	4×4	5×4	6×4	7×4	8×4	9×4	10×4
1×3	2×3	3×3	4×3	5×3	6×3	7×3	8×3	9×3	*10×3*
1×2	2×2	3×2	4×2	5×2	6×2	7×2	8×2	*9×2*	10×2
1×1	2×1	3×1	4×1	5×1	6×1	7×1	8×1	9×1	10×1

Note: The thicker line in the table separates square M11 from square M22 in the multiplication matrix. Examples of tasks included in the interview appear in italics.

Table 6 *The table shows the division of the multiplication facts that was the basis for choosing interview tasks.*

Only a few combinations in multiplication were automated. This was true for multiplication with 1, 5 and 10 as well as a few other scattered tasks. Otherwise, Daniel used to count upward. In certain problems, he counted upward from the beginning, and for others counted upward from a known product. This upward counting occurred occasionally, with one of the factors as a step, occasionally with the number 1 as a step, and occasionally with both variations within the same task. Daniel's way of thinking can be summarized in the following way:

I. *Counting up by the factor on the left from an automated product* (in one or several steps)
Example 7×9: "56, 63"

II. *Counting up by 1s from an automated product*
Example 7×8: "42, 43, ..., 49, 50, ..., 56"

III. *Counting up by the factor on the left from the beginning*
Example 3×5: "3, 6, 9, 12, 15"

IV. *Counting up by the factor on the left from the beginning or from an automated product. Transfer to counting up by 1s*
Example 6×8: "6, 12, 18, 24, 25, 26, ..., 30, 31, ..., 36, 37, ..., 42, 43, ..., 48"

Daniel had mastered the distributive law, in the sense that he was able to break down a task such as 7×8 into (7×6)+(7×2). When he did not find an automated combination that contained the first factor near the product he sought, he chose, instead, to begin from the beginning. When he was unsure of the size of the factor for counting up, as in some cases of crossing over to a new column, he chose to regress to counting up by 1s. When he counted up by one factor, he had only to use two parallel counting sequences. However, when he counted up by 1s, as he did in the example 6×8, it was necessary for him to apply (from 25 upwards) *three parallel counting lines*:

24	25	26	27	28	29	30	31	32	33	34	35	36	37	38	39	40	41	42	43	44	45	46	47	48
	1	2	3	4	5	6	1	2	3	4	5	6	1	2	3	4	5	6	1	2	3	4	5	6
4						5						6						7						8

While he mumbled out the numerals in the upper counting sequence, he used his fingers repeatedly from 1 to 6. He also used his fingers for the third sequence, but only up to 4. At this point, "external counters" were needed for counting upwards from 1 to 6. Thus he then used a "pattern" of 8 books on a bookshelf nearby as a replacement for his fingers, which were already in use. With the help of the books, he kept track of the other factor from 4 to 8.

Daniel also admitted that he used Italian numerals for such counting sequences as 6, 12, 18, 24, 30, 36, 42, 48. He was born in Sweden of Swedish parents but explained that he had gone to school in Italy during the 2nd, 3rd, and 4th Grades and that he had learned this way of thinking there. Thus, altogether, Daniel used three parallel sequences of counting up to number 48, of which one was in Italian and the other two in Swedish! When his teacher asked, at the end of the interview, why he had never said that before (he was then in the 8th Grade), Daniel answered:

"You never asked!"

Daniel is very dependent upon understanding multiplication as repeated addition "from the right". Thus 7×8 means (7+7+7+7+7+7+7+7) and not (8+8+8+8+8+8+8). In this sense, the commutative law for multiplication does not function for Daniel. He is not immediately helped by 5×8 as a support for 6×8, which one often takes for granted. Instead 6×7, or perhaps even more, 6×6 would be appropriate as a support if one is to pursue Daniel's way of thinking further.

Daniel never used as a memory support a product which is larger than the product he was seeking. When asked directly on 9×6, he answered the he never thought of using 10×6. He is happy to have gotten that tip, and I believe it was a good support for his continued work. The interview on subtraction shows that Daniel was very sure of arithmetic tasks of the type 10-6 and should therefore be able to manage the subtraction necessary to arrive at 54 from 60.

The interview with Daniel reveals why it took him so much time to give answers to what one may think are simple multiplication facts and simple tasks using the multiplication algorithm, and why his answers so often were wrong. His ways of thinking have also become consolidated over many years of skills training, which made it extremely difficult to bring about the qualitative changes in his thinking. Further training of skills, without Daniel's first having had the opportunity to develop his way of thinking, would in all probability only further consolidate the primitive and lengthy method observed during the interview. Instead, it is necessary to look for qualities that are more easily developed and which, in fact, Daniel has, and then try to lead him out of the circle in which he is stuck. That would also help him to acquire a deeper insight into the meaning of the division concept and develop good skills in this area. Division posed a problem for Daniel, as can be seen in the results of the written test, that are even greater than the difficulties he had in multiplication.

Division

The tasks in the interview on division were chosen, as mentioned earlier, from the groups corresponding to those used for multiplication. Daniel had only automated a few combinations. In other tasks, he moved forward in the same way as for multiplication. It is sufficient here to describe his method. Daniel demonstrated how he will go about answering the problem 48/6. He arrives at the answer after about 20 seconds, and his answer was:

"48/6=6"

As in the corresponding multiplication tasks, Daniel uses three parallel sequences of counting on the way to his answer.

And thus it was not particularly surprising that he missed the two final stages, 7 and 8, in the sequence farthest down. This time, his working memory did not stretch far enough to keep track of all the steps — the same explanation for most of the incorrect answers he gave to problems in the division test.

Because of Daniel's understanding of division, he had an easier time with problems with larger denominators than those with smaller ones. In the task 16/8, he needs only to count up for two steps (8, 16), while in the task 16/2 he must count eight steps (2, 4, 6, 8, 10, 12, 14, 16). He had difficulty seeing that 16/2 becomes 8 by going trough 8, 16 or 8×2 and therefore seldom fluctuated between both aspects of division when the one way of thinking was simpler. Compare with "choice" in subtraction (Carpenter & Moser, 1984) and with "content and partition division" (Wigforss, 1925). Even in a case where the denominator was as small as 3, as in 27/3, Daniel was completely dependent upon counting up using the denominator from the beginning: 3, 6, 9, ..., 27. This is of course a consequence of 9×3 not being a part of his automated combinations as well as his not thinking of the number 27 in terms of a multiplicative "part-part-whole".

7. DIAGNOSTIC AND DIDACTIC THEORY OF CONTENTS

The purpose of this relatively extensive description of an interview with one student is not to show how students with gaps in their knowledge of and skills in arithmetic should be instructed. It is to draw attention to the significance of assessing the students' complex knowledge and skills in mathematics, a necessary complement to other forms of diagnostic assessment.

In the past 15 years, a great number of qualitative studies have been carried out on students' understanding of basic concepts and ideas in mathematics, and on the ways of thinking they use when they calculate and solve problems (Nesher & Kilpatrick, 1990). These studies have helped to give a firm foundation to both the form of development and the content of diagnostic assessment in mathematics.

At the same time, I would like to point out that the research mentioned above must not be limited to investigating students' ways of thinking within the small restricted "islands of subject content". It must also examine students' thinking in larger, content-related, and didactically associated areas such as concept fields (Vergnaud, 1990), different types of problem areas such as I describe in this chapter, and large, interconnected areas in which mathematics is used and applied outside the school. We need to develop better didactical *theories of content* on the areas which our students must learn. These theories should provide a foundation for the selection and organization of content and activities, not only for diagnostic assessment, but also for instruction and learning in mathematics.

I believe that individual teachers should, at all times and with interest, curiosity, and respect for students' thinking, successively try to develop and deepen their students' complex knowledge of and skills in mathematics. In this way, the teachers become more skilled in assessing the results of their work and of their students' learning. It will also become easier to understand learning as a phenomenon. What it is possible or reasonable for a teacher to learn in teacher education, continued teacher education, or practical professional activities must, however, be an object of further study.

NOTE

I am very grateful to Jeremy Kilpatrick, University of Georgia, and to the editor Mogens Niss, for helpful comments and editorial work on earlier drafts.

REFERENCES

Brousseau, G.: 1984, *The Crucial Role of The Didactical Contract in the Analysis and Construction of Situations in Teaching and Learning Mathematics* Occasional Paper 54, IDM, Universität Bielefeld.

Carpenter, T.P. & Moser, J.M.: 1984, 'The acquisition of addition and subtraction concepts in grades one through three', *Journal for Research in Mathematics Education* **13**(a), 179–202.

Johansson, B.: 1975, *Aritmetikundervisning*, PUMP-projektet nr 9, Pedagogiska Institutionen, Göteborgs Universitet.

Johansson, B. & Kilborn, W.: 1974, *Ett diagnosinstrument för aritmetikundervisning*, PUMP-projektet nr 5., Pedagogiska Institutionen, Göteborgs Universitet.

Johansson, B.: 1986, 'Kunskaper om kunskaper och färdigheter i matematik' [Knowledge of and skills in mathematics], Institutionen för Ämnesdidaktik, Göteborgs Universitet.

Johansson, B. & Kilborn, W.: 1982, *Räkning*, Liber Utbildningsforlaget, Stockholm.

Kilborn, W.: 1979, *PUMP-projektet. Bakgrund och erfarenheter* [Background and Results], Liber Utbildningsforlaget, Stockholm.

Lgr 80: 1980, *Läroplan för grundskolan. Allmän del*, Liber Utbildningsforlaget, Stockholm.

Nesher, P. & Kilpatrick, J. (Eds.): 1990, *Mathematics and cognition: A research synthesis by the International Group for the Psychology of Mathematics Education* (ICMI Study Series), Cambridge University Press, Cambridge.

Vergnaud, G.: 1990, 'Epistemology and psychology of mathematics education', in Nesher, P. & Kilpatrick, J. (Eds.), *Mathematics and cognition: A research synthesis by the*

International Group for the Psychology of Mathematics Education (ICMI Study Series), Cambridge University Press, Cambridge.

Wigforss, F.: 1925, *Den grundläggande matematikundervisningen* [Basics in Teaching Mathematics], Magnus Bergvalls förlag, Stockholm.

Wigforss, F.: 1931, 'Rostads standardprov i sifferräkning', in *Rostads elevförbunds årsskrift 1931*, 86–100, Rostads elevförbund, Kalmar.

Wigforss, F.: 1934, 'Rostads standardprov och tabeller i problemlösning', in *Särtryck ur Rostads elevförbunds årsskrift 1934*, 3–9, Rostads elevförbund, Kalmar.

Wigforss, F.: 1937, 'Hur reda sig barnen med additionsuppgifterna inom talområdet 1–10?', in *Särtryck ur Rostads elevförbunds årsskrift 1937*, 14–32, Rostads elevförbund, Kalmar.

Bengt Johansson
Institutionen for Ämnesdidaktik,
University of Gothenburg,
Sweden

JOHN IZARD

CHALLENGES TO THE IMPROVEMENT OF ASSESSMENT PRACTICE

1. INTRODUCTION: THE ROLE OF ASSESSMENT

Assessment has the function of providing valid evidence of learning achievement to inform students, to facilitate provision of further learning, or to certify that a required level has been reached. Such information is of particular relevance to individual students and their teachers. Teachers are able to develop and improve the educational process if they have identified the strengths of their pupils and know which areas of study require attention.

"The information provided by assessment should do more than portray a learner's level of performance. It should inform the actions of all participants in the learning situation. ... Links must be forged between the assessment, the instruction it reviews, and the instruction it anticipates" (Clarke, 1989, p. 4).

Assessment procedures used by examining authorities and within schools have marked influence on the instruction carried out in schools. Crooks (1988) carried out a comprehensive survey on the effects of classroom evaluations on students, summarizing results from 14 specific fields of research. He noted that formal testing under carefully controlled conditions may be only a small component of the evaluation activities in a particular class, but argued that other forms of evaluation such as teacher questioning, commenting on classroom performance, and use of checklists had similar effects to tests and test-like activities. He listed effects of evaluation on students at the lesson level, the course or learning module level, and over the longer term.

Gasking (1948) argues that external examinations exert considerable pressure on the subjects taught, on the topics within those subjects, and on the teaching strategies which are used. He draws two conclusions:

"First, a reformer, a teacher or an administrator, who wishes to introduce certain changes into school education, will, if the examinations remain the same, find his efforts defeated in the long run, unless the changes he introduces increase the chances of examination success" (p. 10).

M. Niss (ed.), Investigations into Assessment in Mathematics Education, 185–194.

"But conversely , if a reformer succeeds in introducing changes in the type of examination, he will automatically tend to bring about such changes in the content and method of education as are likely to make for success in the changed type of examination" (p. 10).

Gasking also discusses the effects of indirect measurement as compared with the direct measurement of relevant skills and knowledge.

"The easiest way...for teachers to get good results is to concentrate on imparting those capacities which are directly measured in the examination, even when these are not the real objectives of the course, but are merely capacities which the examiners are trying to use as indirect measures. Thus education is perverted away from its real objectives, on to that which examiners use as indirect measures" (p. 14).

Since the easiest things to measure relatively objectively are knowledge of facts (particularly isolated facts rather than interrelationships) and knowledge of the verbal expression of theories, generalizations and definitions, examinations tend to have high proportions of such questions, rather than questions which test understanding, or applications in novel circumstances. Initially, using factual recall provides an indirect measure of the higher-order skills based on the assumption of a reasonably high correlation between the intended skills and those actually measured. However intensive coaching in the indirect criterion tends to reduce this correlation, and the indirect measure ceases to be a valid indicator of the true objectives (Gasking, 1948, p. 15).

Gasking points out that when examiners perceive that students without the intended skills are passing they may raise the standard required in the indirect criterion. This diverts schools even more from the intended skills and makes the examination an even less valid measure of those skills. The only solution is to devise direct measures.

Madaus (1988) reaches similar conclusions when he summarizes the considerable literature on the power of testing practices to influence curriculum, teaching and learning. When students, teachers and administrators believe that the results of an examination are important, their actions are guided by this perceived importance. Where measurement of learning achievement (or any other social indicator) is used more frequently for important social decisions, the social progresses being monitored are more likely to be distorted by teachers teaching to the test, by the development of test-taking skills rather than better teaching, and by the tendency to treat test results as the major goal rather than the achievement itself. The claim that narrow testing practices have undesirable influences on learning and teaching in schools (Palincsar & Winn, 1990) receives considerable support from educators, educational researchers, and cognitive psychologists (see for example, Batten, 1989; Garcia, Rasmussen, Stobbe & Garcia, 1990; Haney & Madaus, 1989; Hatch & Gardner, 1990; Neill & Medina, 1989; Smith & Klass, 1989; Withers & McCurry, 1990).

Students come to believe that the study of mathematics is the development of skills that are required by their teacher (examination board) under test conditions. If they can do well on such tasks, they believe that they are skilled in mathematics. This influence can be seen at all levels of schooling, and is perpetuated by those who are proficient in mathematics and who become teachers of mathematics in turn.

2. MEETING THE CHALLENGE OF APPROPRIATE ASSESSMENT

The assessment procedures implicitly define the skills regarded as important by the examining authorities and schools.

"Even when the precise examination questions are not known by the teachers, the format is generally standardized. Thus it is not uncommon for schools to give their students considerable practice in answering questions that are framed in the format of the test rather than in more natural contexts. Students may be given vocabulary words and their definitions to memorize, ignoring the fact that the meaning of language comes from its context. Thus, the learning will be artificial and designed to do well on examinations rather than in showing students how to deduce meaning from context" (Windham, Levin & Bathory, 1990, p. 85).

As Somerset & Ekholm (1990, pp. 27–28) have noted, where examination centers have limited access to experienced examiners and associated resources the task of preparing questions which reflect the intentions of curriculum is made more difficult. In order to meet deadlines for production of examination papers under such constraints the tendency is to choose types of questions which can be produced quickly rather than those which sample the range of skills proposed in the curriculum. Such examination papers have a substantial proportion of questions devoted to recall of learned material directly from memory, rather than questions which assess the ability to *apply knowledge in new situations*, to *interpret information in context*, to *relate causes, consequences and reasons* or to develop a sequence of steps to *make a decision* or to *solve a problem*. One serious consequence is that only those skills assessed by the examination are taught and studied rather than the range of competencies required by the syllabus.

The assessment of mathematics has to take into account the requirements of sound assessment. Professional test constructors work to a set of specifications to ensure that all of the questions are *relevant*, that the questions are *representative* to the tasks that the curriculum sets out as appropriate, that there is a *balance* between the various topics, and that the *emphasis* in the curriculum is reflected in the choice of questions for the test (Izard, 1990). Traditional examinations, with their emphasis on performance within a restricted time-frame and the other constraints of pencil-and-paper testing (whether multiple-choice or open-ended), have meant that traditional assessment procedures address only a portion of the area we would recognize as mathematics.

We expect students to be able to cope with a range of mathematical tasks. To be fair, our selection of tasks for assessment purposes should be representative of the tasks we expect them to master. This implies that the tasks should include problems in appropriate contexts where there are several solutions and different solution strategies, where there are "real" problem situations, and where higher-order thinking can be demonstrated. However, such assessment still has to be practical, reliable and valid, and account has to be taken of the way in which the scores on each task are aggregated in order to report a level of achievement (Izard, 1991a). The assessment procedures that are adopted have to be efficient in gathering representative evidence so that the assessment procedures

- provide useful information about the learner
- minimize incursions into teaching time
- minimize the teacher's load, and
- relate to all significant curriculum intentions.

Such procedures have to be representative so that reasonable inferences can be made about the learning achievement that was *not* assessed.

Assessment strategies may go beyond traditional subject boundaries. For example, Clements (1980, 1982) has reported on Australian studies investigating the errors made by children on written mathematical tasks. The investigations of errors included the Newman interview procedure (Newman, 1977, 1983) which allows the type of error to be classified. These classifications are shown below in Table 1. These studies show the crucial role of language in mathematical problem solving. Marinas & Clements (1990) have applied the Newman interview procedure in Malaysian secondary school.

3. ASSESSMENT AND THE TEACHER

Black & Dockrell (1984) recognized that teachers require extensive inservice work on assessment techniques if improved assessment strategies are to be implemented. This view is supported in the literature (for example, Hoge & Coladarci, 1989; Nitko, 1989; Omar Hashim, 1990; Brown, 1990b) and is consistent with the view expressed by Brandsma, ten Brummelhuis, & Plomp (1989) with respect to implementing innovative instructional materials.

In addition to developing a wider use of assessment strategies, teachers need procedures for combining or aggregating scores so that one component does not exert undue influence (Izard, 1991a). To interpret combined scores a number of assumptions must be made. The assessment tasks are assumed to be relevant, and sampled from the knowledge and skills required by the student in his or her study of the subject. Further, it is

Reading errors. An error would be classified as if the student could not read a key word or symbol in the written problem to the extent that this prevented him/her from proceeding further along an appropriate problem-solving path.

Comprehension errors. The student had been able to read all the words in the question, but had not grasped the overall meaning of the words and, therefore, was unable to proceed further along an appropriate problem-solving path.

Transformation errors. The student had understood what the question wanted him/her to find out but was unable to identify the operation, or sequence of operations, needed to solve the problem.

Process skills errors. The student identified an appropriate operation, or sequence of operations, but did not know the procedures necessary to carry out these operations accurately.

Encoding errors. The student correctly worked out the solution to a problem, but could not express this solution in an acceptable written form.

Table 1 *Newman classification of errors (Newman, 1977, 1983)*

assumed that the sample of tasks is representative of all such tasks and that the sample reflects the balance of such tasks.

There are several perspectives to consider when combining marks. One perspective denies meaning to a combined score unless there is internal consistency between the various tasks, at least with respect to intercorrelations between scores on those tasks. In a test-development context, researchers use procedures such as reviewing discrimination indices and item characteristic curves and factor analysis when deciding whether a question "belongs" to the test or not. From this perspective questions judged inappropriate are deleted from the test. However deleting questions on this basis can result in the final form of the test being less representative than specified originally. From another perspective, the meaning of combined scores is defined arbitrarily (by teacher or examination board), regardless of the patterns of correlation between scores on the tasks. The combination of such scores to arrive at a final mark in a mathematics subject may lead to unintended outcomes. It cannot be assumed that simple additions of raw scores from separate mathematics assessment tasks such as marks for *class work*, *study projects*, and *examination questions* leads to combined scores with the intended weighing of components. To extend an analogy with money presented by Viviani (1990), a score of three obtained by combining one US dollar with two Sterling pounds will have a different value to a score of three obtained by combining two US dollars with one Sterling pound. We need to know the relationship between the currencies and to express each in some comparable way so that combined scores that are similar represent similar worth. As McGaw et al. (1990, p. 30) have observed,

"The aggregation of unscaled marks is inappropriate and unjustifiable. If individual subjects are too different for scaling of their results on to a common scale to make sense, then it makes no sense to add the results. Another way to make this point is to say that whenever it is believed appropriate to add results then it is appropriate to express them on a common scale before doing so."

4. ENGAGING THE STUDENT IN THE ASSESSMENT TASK

Earlier in this paper the need for assessment procedures which gather representative evidence was emphasized, particularly with respect to all significant curriculum intentions. There is another important perspective on this notion of representative evidence. The use of assessment tasks in examinations and school-based assessments for inferences about student achievement levels rests on an assumption that each student was prepared to participate in providing evidence by tackling the task. Clearly, results for a student lacking in motivation to engage in such tasks may not provide a valid measure of the achievement of that student.

Improved assessment practices must take account of both student perceptions of what is relevant in the study of mathematics, and student willingness to engage in individual and/or group tasks which differ from traditional tasks. Mathematics students studying for traditional external examinations may resent the introduction of more authentic mathematics problems into a syllabus because they see such tasks as wasting time they could otherwise use in preparing for the examinations. Further, since such examinations are competitive, students (and teachers?) may need considerable training to accept that cooperative approaches to solving mathematics problems are worthwhile. If the assessment procedures do not provide credit for individual or group work on authentic problems, then attempts to include such problems into mathematics syllabuses would appear to be doomed.

As well as ensuring that the assessment tasks are representative of the tasks that the curriculum sets out as appropriate, such tasks need to appeal to students. The excitement of challenging mental tasks and the pleasure and satisfaction of achieving success on such tasks are important in encouraging students to continue their study of mathematics. When you see students playing computer games you gain some idea of the power of motivation. Hours of voluntary effort are spent striving to improve on previous scores. Perhaps we should be trying harder to harness such influences to assist the learning of mathematics.

5. CHOOSING MORE DIVERSE ASSESSMENT TASKS

The selection of tasks for assessment purposes (or the extraction of assessment information from teaching tasks) requires comprehensive knowledge of the intended outcomes of a mathematics syllabus and recognition that the actual outcomes will be a distorted version of the intended outcomes if the assessment tasks do not match the intended outcomes. Given Gasking's warnings discussed earlier in this paper, such assessment tasks should be direct rather than indirect wherever possible. A number of direct tasks which have been tried in Australian classrooms are now described.

At primary or junior secondary level *puzzle tasks* can be constructed using tangram pieces, tetrominoes, pentominoes cut from light card or stiff paper. For each candidate the test sheet can present a variety of regions based on combinations of the pieces but the particular pieces required are not identified. The task is to choose the appropriate pieces from those provided and to assemble them to make the required figure. Successful completion may be recorded by the teacher as each task is done or students may use a glue-stick to fasten their chosen pieces in place.

Junior secondary students have been given three-dimensional puzzles to solve. Using the pieces of the Danish puzzle know as the *Soma Cube*, students were required to use some of the seven pieces to assemble given solids. Students were not told which pieces were required. Many male and female students in their ninth year of schooling had trouble in completing these tasks, while others in their sixth year of schooling could complete such puzzles in less than a minute (Izard, 1989). (If a puzzle had not been solved in five minutes I suggested that they tackle another puzzle).

Those teachers who have access to computers can present tasks on the screen, with the program recording the response made, the time to make the response, and (if desired) providing immediate feedback to the student as to the correctness of the response. My research has included presentation of *spatial* (embedded figures) *tasks* to students in the seventh, eighth, and ninth years of schooling using laptop computers (Izard, 1989). The most attractive features of such assessment tasks are that students have all the time they need to do each question, the scoring is done automatically, and information is provided on every question (rather than just those questions that were finished in a fixed time).

Unlike their experiences with many mathematics tests, students find these types of test enjoyable. Such tests invite student comments like, "This is fun. Why can't our other tests be like this?" and "Do we have to go back to the mathematics class? Haven't you got any more tests like this?". These tasks appeal to students because each task differs from the usual test question, demands mathematical skills other than computation, may involve alternative correct answers (just as in the "real" world), and in many cases students know immediately when a correct solution is achieved. In validity terms the tasks are more direct (as the term is used above by Gasking), and therefore more likely to retain validity if teachers teach to the test.

6. CONCLUSION

These approaches have been used successfully in Australian classrooms to broaden the range of assessment tasks, and to change the emphasis to test where "time to complete" was the criterion rather than "how many questions in fixed time". Students become more engaged in the tasks than is customary for traditional examinations, have a much better idea of how they performed in the test situation, and some students whose mathematical skills are not recognized in traditional examinations can perform at a sophisticated level. Further, the teacher can observe a wider range of student responses and each student's view of mathematics in enhanced.

The choice of more diverse assessment tasks must still meet the requirement that assessment has to be practical, reliable, and valid (Izard, 1991b). This applies regardless of whether the assessment is at primary, secondary, or tertiary level. Test analyses of the Australian data for the practical tasks using the Soma Cube pieces and the computer administered embedded figures tasks show that such assessments have similar psychometric properties to traditional examinations (Izard, 1989). Similar results are obtained in analyses of marking schemes for projects in mathematics at university and polytechnic level (Izard & Haines, 1991). Further, analyses show that results on these tasks can be combined with traditional results in a meaningful way.

There are many challenges to the improvement of assessment practice with respect to mathematics education. Traditional open-ended and/or multiple-choice examinations fail to assess many important aspects of mathematics which are intended curriculum outcomes but traditional examinations are powerful influences in directing teaching and learning. Students are constructive participants in learning, relating new events to prior knowledge and understanding, but traditional examinations are not well suited to exploring conceptual knowledge in "real life" contexts (Gunstone, 1987; Adams, Doig & Rosier, 1991). For traditional examinations the research literature on item analysis, test reliability and validity is extensive. Alternative assessment approaches which allow other significant curriculum outcomes to be monitored are required, and these approaches should be supported by research. To use new strategies without gathering evidence of validity (implying reliability as well) would be counter-productive in seeking to improve mathematics education.

REFERENCES

Adams, R.J., Doig, B.A. & Rosier, M.J.: 1991, 'Science learning in Victorian schools', *ACER Research Monograph No. 41*, ACER, Hawthorn, Victoria.

Batten, M.: 1989, 'Year 12: Student's expectations and experiences', *ACER Research Monograph No. 3*, ACER, Hawthorn, Victoria.

Black, H.D. & Dockrell, W.B.: 1984, *Criterion-referenced assessment in the classroom*, Scottish Council for Research in Education, Edinburgh, UK.

Brandsma, J., ten Brummelhuis, A. & Plomp, T.: 1989, 'Evaluation of national curriculum and implementation in junior vocational education', *Studies in Educational Evaluation*, **15**, 359–382.

Brown, M.: 1990, 'Problematic issues in national assessment', paper presented at the Education Reform Act Research Network Seminar at the University of Warwick, 28 June 1990.

Clark, D.J.: 1989, *Mathematics Curriculum and Teaching Program (MCTP): alternatives in mathematics*, Curriculum Development Centre, Canberra, A.C.T.

Clements, M.A.: 1980, 'Analyzing children's errors on written mathematical tasks', *Educational Studies in Mathematics*, **11**, 1–21.

Clements, M.A.: 1982, 'Careless errors made by sixth-grade children on written mathematical tasks', *Journal for Research in Mathematics Education*, **13**, 136–144.

Crooks, T.J.: 1988, 'The impact of classroom evaluation practices on students', *Review of Educational Research*, **58**, 438–481.

Garcia, E., Rasmussen, B., Stobbe, C. & Gracia, E.: 1990, 'Portfolios: Assessment tool in support of instruction', *International Journal of Educational Research*, **14**, 431–436.

Gasking, D.A.T.: 1948, *Examinations and the aims of education*, Melbourne University Press, Carlton, Victoria.

Gunstone, R.F.: 1987, 'Student understanding in mechanics: A large population Survey', *American Journal of Physics*, **55**, 691–696.

Haney, W. & Madaus, G.: 1989, 'Searching for alternatives to standardized tests; Whys, whats and whithers', *Phi Delta Kappan*, **70**, 683–687.

Hatch, T. & Gardner, H.: 1990, 'If Binet had looked beyond the classroom: The assessment of multiple intelligences', *International Journal of Educational Research*, **14**, 415–429.

Hoge, R.D. & Coladarci, T.: 1989, 'Teacher-based judgements of academic achievement: A review of the literature', *Review of Educational Research*, **59**, 297–313.

Izard, J.: 1989, 'Modelling student performance on assessment tasks involving three-dimensional models', paper presented at ICTMA-4, Roskilde, Denmark, July, 1989.

Izard, J.: 1990, 'Assessing achievement with non-pencil-and-paper tasks', paper presented at the Fifth South East Asian Conference on Mathematics Education, Bandar Seri Begawan, Brunei Daussalem, 14–17 June, 1990.

Izard, J.: 1991a, 'Issues in the assessment of non-objective and objective examination tasks', in Luijten, A.J.M., *Issues in public examinations*, Lemma, B.V., Utrecht, The Netherlands.

Izard, J.: 1991b, 'Current developments in assessment of mathematical projects', invited paper presented at the City University/Polytechnic South West (Plymouth) Workshop on assessment of Mathematical Projects, Exeter University, April, 1991.

Izard, J. & Haines, C.: 1991, 'Making schemes for projects in mathematics and its applications: Interpreting the results', paper presented at ICTMA-5, Utrecht, The Netherlands, September, 1991.

Madaus, G.F.: 1988, 'The influence of testing on the curriculum', in Tanner, L.N. (Ed.), *Critical issues in curriculum, Eighty-seventh Yearbook of the National Society for the Study of Education* Part I, 83–121, National Society for the Study of Education, Chicago, IL.

Marinas, B. & Clements, M.A.: 1990, 'Understanding the problem: A prerequisite to problem solving in mathematics', *Journal of Science and Mathematics Education in South East Asia*, **13**, 14–20.

McGaw, B., Eyers, V., Montgomery, J., Nicholls, B. & Poole, M.: 1990, *Assessment in the Victorian Certificate of Education: Report of a review commissioned by the Victorian Minister for Education and the Victorian Curriculum and Assessment Board*, Australia: Victorian Curriculum and Assessment Board, Melbourne, Victoria.

Neill, D.M. & Medina, N.J.: 1989, 'Standardized testing: Harmful to educational health', *Phi Delta Kappan*, **70**, 703–713.

Newman, M.A.: 1977, 'An analysis of sixth-grade pupils' errors on written mathematical tasks', in Clements, M.A. & Foyster, J. (Eds.), *Research in mathematics education in Australia 1977* (Vol. 1, 239–258), Swinburne College Press, Melbourne, Victoria.

Newman, M.A.: 1983, *The Newman language of mathematics kit*, Harcourt Brace Jovanovich, Sydney.

Nitko, A.J.: 1989, 'Designing tests that are integrated with instruction', in Linn, R.L. (Ed.), *Educational Measurement* (3rd Edition), 447–474, American Council on Education & Macmillan, New York.

Omar Hashim: 1990, 'The balance between external and internal assessment', paper presented at the Sixteenth International Conference, International Association for Educational Assessment, Maastricht, The Netherlands, 18–22, 1990.

Palincsa, A.S. & Winn, J.: 1990, 'Assessment models focussed on new conceptions of achievement and reasoning', *International Journal of Educational Research*, **14**, 411–413.

Smith, L.M. & Klass, C.S.: 1989, 'Assessing the whole child', in Simons, H. & Elliott, J. (Eds.): *Rethinking appraisal and assessment* (Chapter 12), Open University Press, Milton Keynes.

Somerset, A. & Ekholm, M.: 1990, 'Different information requirements for different levels of decision-making', in Ross, K.N. & Mählck, L. (Eds.), *Planning the quality of education: The collection and use of data for informed decision-making*, 15–33, UNESCO, Paris, and Pergamon, Oxford.

Viviani, N.: 1990, *The review of tertiary entrance in Queensland 1990*, (Report submitted to the Minister for Education), Department of Education, Brisbane, Queensland.

Windham, D., Levin, H. & Bathory, Z.: 1990, 'Issues in the design and development of educational information systems', in Ross, K.N. & Mählck, L. (Eds.), *Planning the quality of education: The collection and use of data for informed decision-making*, 79–95, UNESCO, Paris, and Pergamon, Oxford.

Withers, G. & McCurry, D.: 1990, 'Student participation in assessment in a co-operative climate', in Low, B. & Withers, G. (Eds.), *Developments in school and public assessment*, 77–101, (Australian Education Review, No. 31), ACER, Hawthorn, Victoria.

John Izard
Australian Council for Educational Research
Hawthorn, Victoria,
Australia

MALCOLM SWAN

IMPROVING THE DESIGN AND BALANCE OF MATHEMATICAL ASSESSMENT

1. INTRODUCTION

In recent years, considerable attention has been paid to improving the range and balance of activities that take place in mathematics classrooms. More emphasis is being given to the tackling of practical, extended tasks, which develop higher order skills such as the ability to formulate a question in mathematical terms, select appropriate techniques, and interpret and communicate mathematical results. These aspects are now widely recognized as important and are described in documents around the world (MSEB, 1989, 1990; NCTM, 1989; CITMS, 1982; DES, 1989; NCC, 1989).

The implementation of these curriculum developments is, however, inhibited by most current assessment practices. We continue to assess isolated mathematical techniques through short, closed, stereotyped questions set in pure or artificial contexts. We set such questions because we find them convenient and reliable to mark, but the backwash effect on the curriculum is often disastrous. Such examination questions are bound to encourage imitative rehearsal and practice on similar tasks in the classroom.

The concern of this chapter, therefore, is to explore ways in which we might broaden our assessment practices to include those aspects of mathematical performance which we value. We briefly consider the purposes of assessment, the kind of curriculum we are aiming for, and the design of assessment tasks and marking schemes which might support such a curriculum.

2. THE PURPOSE OF ASSESSMENT

Three broad types of assessment are frequently identified:

- *Formative assessment* which recognizes the achievements of a pupil, so that appropriate follow-up may be designed and provided. Here we include *diagnostic assessment*, where learning difficulties and misconceptions are identified. Such assessment precedes and directs teaching and is designed mainly for the benefit of pupils.

M. Niss (ed.), Investigations into Assessment in Mathematics Education, 195–216.

- *Summative assessment* which measures and records overall achievement in a systematic manner. This usually occurs towards the end of a course and is designed primarily to inform teachers or employers, perhaps to aid selection procedures.
- *Evaluative assessment* which involves evaluating and reporting on the work of a teacher, school, textbook scheme or any other discrete part of the educational service. This form of assessment is mainly designed for parents or education providers.

Gipps (1990) notes two other purposes which have become particularly relevant in the last ten years, *pupil motivation* and *curriculum control*. Previously, summative, norm-referenced examinations often presented many pupils with tasks which they were likely to fail. This was recognized in the Cockcroft Report (CITMS, 1982), where two "fundamental principles" were stated:

"The first is that the examination papers and other methods of assessment which are used should be such that they enable candidates to demonstrate what they know, rather than what they do not know. The second is that the examinations should not undermine the confidence of those who attempt them".

One possible response to this is to develop a tiered system of *differentiated papers*. Students of higher attainment tackle harder papers than those with lower attainment. With sufficient tiers, papers may be designed to ensure that every candidate is faced with a range of appropriate challenges, a significant proportion of which are achieveable. This will ensure that the examination should be a positive educational experience, and that all students are given an opportunity to show what they know, understand and can do.

This approach, however, requires that a limit must be placed on the grades that are obtained on each of these papers. If all grades were made available on the highest level paper, for example, then on this paper, the lowest grades would again tend to be awarded for evidence of very little achievement, and lower attaining candidates would again be faced with many inappropriate tasks. Similarly, upper grade boundaries must be placed on lower level papers. As Gipps (1990) noted, grade limits placed on papers often *harm* motivation.

A further development with the intention of crediting positive achievement, is the movement away from norm-referenced towards criterion-referenced assessment. Instead of ranking students we look for evidence of particular proficiencies and report these in the form of a profile of achievement. This approach is essential if the results of assessment are to be used formatively. The criteria are usually based on hierarchically-ordered lists of behavioral objectives, developed from research. Figure 1 illustrates this with criteria drawn from the current National Curriculum in

England. Criteria may also recognize achievements beyond the academic, hereby increasing the amount of success experienced by students. For a critical analysis and pragmatic defense of hierarchical criterion-referenced assessment see Noss et al. (1989) and Brown (1989).

Sample criteria from the National Curriculum in England (DES, 1991)

Currently, the National Curriculum in England contains approximately 150 criteria under 5 "Attainment Target" headings, arranged into 10 hierarchical levels. The examples below are from Attainment Target 2: Number.

Pupils should be able to:

Level 2

a) Demonstrate that they know and can use number facts, including addition and subtraction.

b) Solve whole number problems involving addition and subtraction.

c) Identify halves and quarters.

d) Recognize the need for standard units of measurement.

Level 4

a) Solve problems without the aid of a calculator, considering the reasonableness of the answer.

b) Demonstrate an understanding of the relationship between place value in whole numbers.

c) Use fractions, decimals or percentages as appropriate to describe situations.

Level 6

a) Calculate with fractions, decimals, percentages or ratio as appropriate.

b) Use estimation to check calculations.

Level 8

a) Calculate with numbers expressed in standard form.

b) Evaluate formulae, including the use of fractions or negative numbers.

c) Understand the significance of approximations.

Level 10

a) Determine the possible effects of errors on calculations.

Figure 1 *Typical criteria from the current English National Curriculum*

There are two major problems associated with criterion-referenced assessment:

Firstly, in order to be meaningful, criteria need to be specific and detailed. The tendency is that criteria proliferate rapidly and fragment the curriculum. The apparent performance of students is increased and the demands on assessors are minimized if the individual elements of the curriculum are taught and tested in isolation. Most worthwhile mathematical tasks, however, involve the deployment of a variety of concepts and skills in more complex situations. The level of difficulty of such tasks is not straightforwardly related to the levels of the individual components. The temptation is, therefore, to leave composite tasks out of the assessment.

Secondly, a detailed hierarchy of criteria, instead of providing a framework for recording attainment, is often misinterpreted as defining the order in which mathematics is *learned*. The criteria actually come to *define* the curriculum and prescribe its delivery. Thus assessment, instead of merely sampling and recording achievement, becomes a method of curriculum control. To illustrate the problems that result, let us suppose a particular criterion is placed at a high level. This may be because research has indicated that few students have achieved it successfully in the past. The reason for this may not be because it concerns an inherently difficult concept, rather it may simply be due to the fact that this concept has been neglected in past teaching practices. If teachers now use the hierarchy to define the order in which they teach, this criterion will be tackled relatively late in students' careers, thus perpetuating the existing state of affairs.

Assessment design should naturally reflect its intended purpose. While diagnostic assessment, for example, may need to focus on narrow criteria, so that particular misconceptions and errors may be unravelled and explored, summative assessment tasks often require a more holistic approach where different threads of mathematics are drawn together. Figure 2, for example, illustrates a useful, but narrow diagnostic task which has been used to discover the extent to which pupils have specific misconceptions concerning place value. This task may be used in a formative way, with remedial action being taken on the basis of the responses. If summative assessments were designed exclusively in this style, however, the curriculum backwash effects would be disastrous.

Which of these numbers has the smallest value?

A	0.625	(34 %)
B	0.25	(3 %)
C	0.375	(2 %)
D	0.125	(37 %)
E	0.5	(22 %)

(APU 1980)

Figure 2 *A diagnostic item. The percentages of 15 year old pupils choosing each response are given in brackets.*

A current pragmatic problem is, therefore, to devise holistic tasks which yield information on a detailed criterion-referenced basis. If we do not succeed, then we effectively encourage teachers to offer mathematics in a fragmented, compartmentalized manner.

To summarize, the design of assessment tasks must reflect the kind of curriculum we want, if we are to avoid unfortunate backwash effects on classroom practice. Our two key principles are therefore:

- *curriculum balance* — each assessment package must consist of a set of tasks of varying length and style which, taken together, reflect our curriculum objectives in a balanced way.
- *curriculum validity* — the assessment tasks themselves must represent learning activities of high educational value so that significant time spent on them will represent a benefit rather than a loss of pupils' learning.

3. WHAT KIND OF CURRICULUM DO WE WANT?

There is now a fairly widespread consensus of the range and balance of mathematical activity that we would ideally like to see in classrooms. This is exemplified by the following list:

- Activities should bring together different areas of mathematics
- The order of activities should be flexible
- Activities should be balanced between tasks which develop knowledge, skills and understanding, and those which develop the ability to tackle practical problems
- Activities should be balanced between the applications of mathematics and ideas which are purely mathematical
- Activities should be balanced between those which are short in duration and those which have scope for development over an extended period
- Activities should, where appropriate, use pupils' own interests or questions either as starting points or as further lines of development
- Activities should, when appropriate, involve both independent and cooperative work
- Tasks should be both of the kind which have an exact result or answer and those which have many possible outcomes
- Activities should be balanced between different modes of learning; doing, observing, talking and listening, discussing with other students, reflecting, drafting, reading and writing, etc.
- Activities should encourage students to use mental arithmetic and to become confident in the use of a range of mathematical tools
- Activities should enable students to communicate their mathematics
- Activities should enable students to develop their personal qualities
- Activities should enable students to develop a positive attitude to mathematics

(NCC, 1989, B8–B11, paragraphs 5.1 to 5.14)

Normal classroom activities often fall a long way short of this ideal. Most popular current mathematics textbook schemes encourage imitative, individualized work through short, closed, tasks. Little opportunity is offered for students to feel what it is like to be a mathematician; formulating questions, pursuing one's own line of inquiry, selecting and choosing the aspects of mathematics to deploy, communicating discoveries, and so on.

The difficulties inherent in using our assessment procedures to encourage such practices are enormous. Typically, we hear: "Examinations are supposed to *reflect* the curriculum, not direct it. We can only assess what teachers are currently teaching" and "It is unfair to set questions this year that are too different from those we had set last year".

Of course, there are ways around these problems by adequately supporting teachers, pre-piloting questions, and so on, but the traditional image that examination questions are successfully written by "lone craftsmen working in their spare time" prevails. The role of an "assessment designer" must receive much greater recognition and status if we are to achieve the quality that is needed.

4. THE DESIGN OF ASSESSMENT TASKS

While designing and evaluating an assessment task, *six dimensions* must be carefully considered:

(i) The achievements being assessed.
(ii) The appropriate method of assessment.
(iii) The degree of "openness" of the task.
(iv) The degree of autonomy and flexibility required.
(v) The extent and coherence of the task.
(vi) The context in which the task is set.

We consider these briefly, in turn, below.

(i) The Achievements Being Assessed

Here, we find it helpful to use the categorization defined in the Cockcroft Report (CITMS, 1982).

Facts and Skills

Facts are items of information which are essentially unconnected or arbitrary. These include notational conventions (e.g.: remembering that (two, three), means the point "2 along and 3 up" on the cartesian plane), and names attached to concepts (e.g.: "origin"). Skills are very well established multistep procedures (the ability to perform long division quickly and accurately).

Facts and skills may be remembered and used with or without understanding the underlying concepts to which they refer.

Conceptual Structures

These are richly connecting networks of concepts and relationships. They make mathematical facts meaningful, underpin the performance of skills, and the presence is shown by the ability to remedy memory failures or adapt procedures to fresh situations. (E.g.: the ability to interpret a graph and translate the information it embodies into alternative representations).

General Strategies

These strategies guide the choice of which skills are appropriate, and enable students to use mathematics to tackle unfamiliar problems. (The processes involved in making choices, planning, distinguishing between important and irrelevant information, classifying, recognizing patterns, generalizing, checking, proving, communicating are all examples of mathematical strategies).

The importance of this dimension to the design of assessment tasks is evident. Whereas a knowledge of facts and skills may be assessed through short closed questions, the existence of conceptual structures and general strategies can only be assessed through more open tasks which require students to make choices, to reason, to explain (see Figures 3 and 4). Indeed, some strategic skills may only be assessable through extended tasks (examples are given later in this paper).

The tasks illustrated in Figures 3 and 4 differ from traditional styles of examination question in that much more emphasis is laid on students reasoning and communicating in non-routine situations.

The marking schemes for "Diagonal Squares" and the "Climbing Game" questions, for example, give credit under the following headings:

- showing an understanding of the problem,
- organizing information systematically,
- describing and explaining the methods used and the results obtained,
- formulating a generalization or rule, in words or algebraically.

The marking scheme for the "Describing Shapes" question assigns full credit (6 points) if;

"the student, using mathematical ideas, effectively gives a coherent description that would result in the reproduction of a correct figure with respect to side length and orientation. If coordinates are used for part (b), the directions for connecting them must be unambiguous. The instructions can be interpreted easily".

The "Swimming Race" and the "Coach Trip" examples require students to interpret and communicate using graphs in realistic contexts. In the former task, it is noticeable how some students interpret the two graphs separately, giving an account of Sheila's progress followed by an account of Jenny's progress. This raises the important issue of whether or not we

Roofs

Roofs can be drawn in different shapes and sizes, using the dots provided.
The first one is 2 3 5 3 .
The second is 3 1 4 1.

(The first number tells you how many units to draw in direction 1, the second in direction 2, the third in direction 3, the fourth in direction 4.)

1. Draw a 2 2 4 2 and a 4 1 5 1.

2. Try to draw a 3 2 5 1 and a 1 4 3 4 .
Explain what happens.

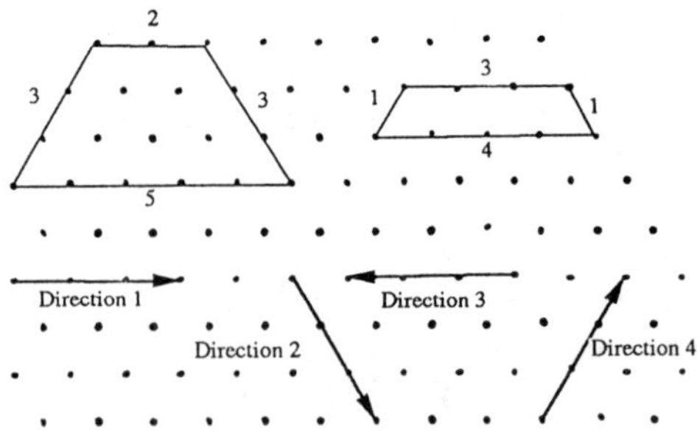

3. Find some rules which enable you to tell, *without drawing*, whether or not you can draw a roof from four given numbers.
Explain why these rules work.

4. If we call the four numbers a, b, c and d respectively, can you express your rules more simply?

Diagonal Square

We will call this shape a 4 x 4 diagonal square.
(Each side is 4 diagonal units long)
There are 25 dots inside this square.

(i) How many dots will lie inside a 20 x 20 diagonal square?
(ii) Explain carefully how you obtained your answer to part (i).
(iii) How would you calculate the number of dots inside a n x n diagonal square?

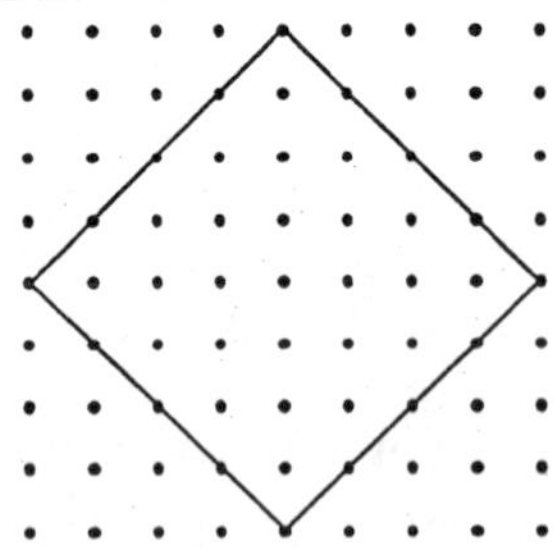

Describing shapes

Imagine you are talking to a student in your class on the telephone and want the student to draw some figures. The other student cannot see the figures. Write a set of directions so that the other student can draw the figures exactly as shown below.

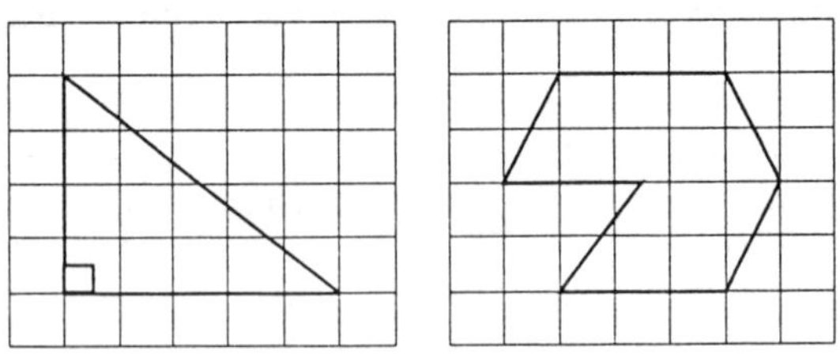

Figure 3 *Tasks designed to assess symbolizing, generalizing and communicating*

The Climbing Game

This game is for two players.
A counter is placed on the dot labelled "start" and the players take it in turns to slide this counter up the dotted grid according to the following rules:

At each turn, the counter can only be moved to an adjacent dot higher than its current position.
Each movement can only take place in one of three directions:

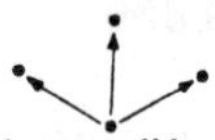

The first player to slide the counter to the point labelled "Finish" wins the game

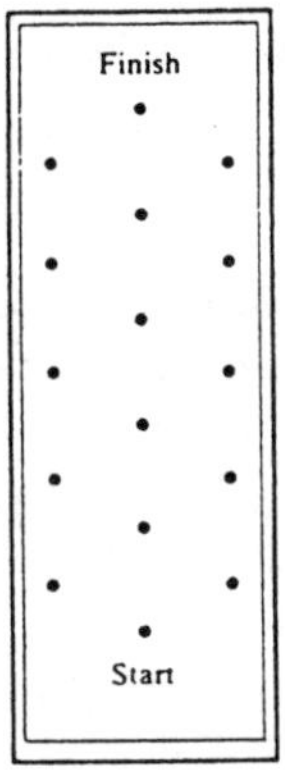

(i) This diagram shows the start of one game, played between Sarah and Paul.
Sarah's moves are indicated by solid arrows. (→)
Paul's moves are indicated by dotted arrows. (- - →)
It is Sarah's turn. She has two possible moves.
Show that from one of these moves Sarah can ensure that she wins, but from the other Paul can ensure that he wins.

(ii) If the game is played from the beginning and Sarah has the first move, then she can always win the game if she plays correctly.
Explain how Sarah should play in order to be sure of winning.

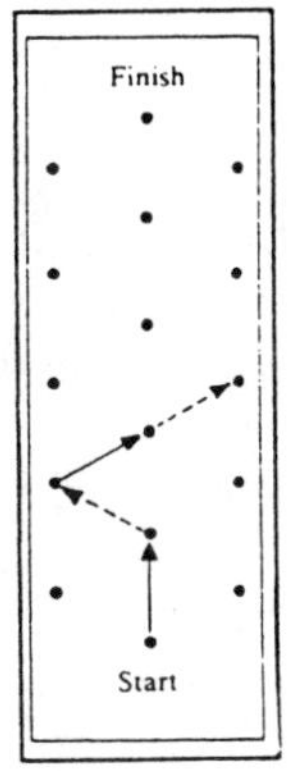

A swimming race

The rough sketch graph describes what happens when two swimmers Sheila and Jenny, enter a 60 metres backstroke swimming race.

Imagine that you are the race commentator.
Describe what is happening as fully and carefully as you can.
You do not need to measure anything accurately.

Distance from shallow end of the swimming pool in metres

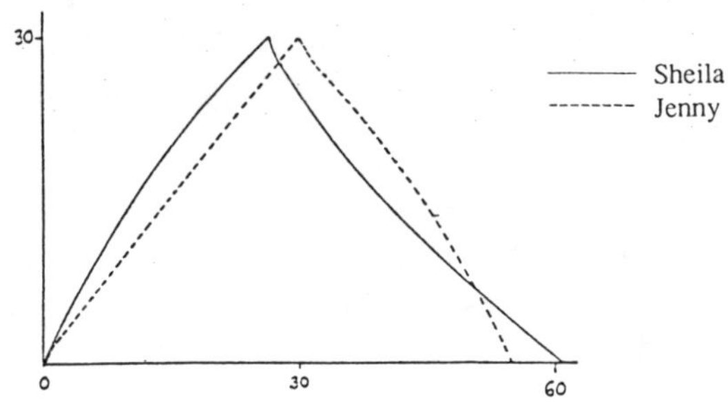

Time in seconds

The Coach Trip

A coach may be hired for £120 per day.
The organiser of the trip decides to charge every member of the group an equal amount for the ride.
(a) Sketch a graph to show how the price of each ticket will vary with the size of the group.
(b) Write down a full explanation for the shape of your graph.

Figure 4 *Tasks designed to assess reasoning, interpreting and communicating.*

assign credit to the quality of writing, as a sports commentary. Perhaps we should if the context is to be taken seriously.

(ii) The Appropriate Method of Assessment

"Purposive" and "incidental" assessment

The assessment of whether a pupil knows, understands, or can use a particular piece of mathematics may be either "purposive" or "incidental". By "purposive" we mean that a task is offered to pupils with the intent that they will use a particular aspect of mathematics. "Incidental" assessment occurs when the teacher spots that the relevant knowledge is being displayed in the normal course of classes. Assessment is thus seen as a *growing collection of evidence.*

While incidental assessment gives a much more reliable indication of achievement, is less intrusive, produces minimal anxiety, and allows more open-ended activity to take place, it does require an efficient and reliable recording system which must be readily internalized by the teacher. In addition, there are considerable practical difficulties in ensuring that each pupil has sufficient opportunity to display his or her capabilities. One method of alleviating these considerable problems is to provide students with versions of criteria and encourage them to inform their teacher when and how they have satisfied particular criteria. This passes some of the responsibility for assessment back to the pupil and encourages reflective activity. While we must welcome any attempt to provide pupils with a clear overview of the intentions of each activity, there remains a danger that satisfying criteria will take precedence over appreciating the activity for its own sake.

Group work, oral and practical tasks

Although it is now widely recognized that an important and effective mode of learning allows pupils to collaborate and discuss in small groups, our methods of assessment usually require pupils to perform in isolation under extreme time pressure. It is questionable whether assessments obtained in this way can yield reliable information on the potential performance of pupils in a future working environment. In addition, the emphasis on purely written modes of response can lead us to seriously underestimate the capabilities of many students. It is therefore necessary to explore ways of assessing individuals within a more natural environment, giving much more emphasis to oral and practical assessment.

(iii) The Degree of "Openness" of the Task

Two dimensions are evident here: the extent to which the student may choose which method to employ in tackling a task, and the number of possible answers or end points that are admissible. Most traditional tasks require students to deploy a given method to obtain a single answer. Much more attention currently is being given to the development of tasks which allow a student to choose which method to deploy and/or which admit a variety of acceptable responses.

As an example, let us consider the following "closed" task.

A number pattern task

1. Complete the table below.

Number of hexagons		Number of matches needed
1	$\longrightarrow$	6
2	$\longrightarrow$	1
3	$\longrightarrow$	...
4	$\longrightarrow$	...
5	$\longrightarrow$	...
12	$\longrightarrow$	...
100	$\longrightarrow$	...

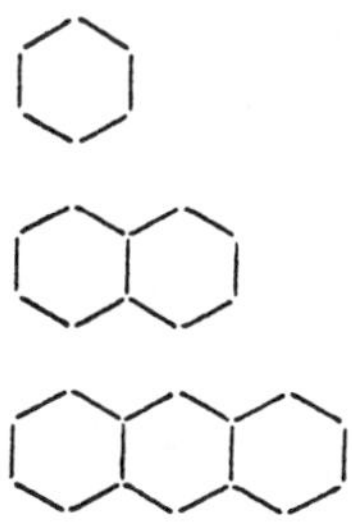

2. Explain in words how to get the second number from the first.
3. Write the mapping in algebra. $h \longrightarrow$

Figure 5 *A "closed" algebra task*

This task could be used to assess whether a student can complete a table or spot a simple pattern and express it using words and algebra. In one experiment, the teacher "opened up" the task by asking students to draw their own shapes on dotted paper and make them grow according to simple rules of their own choosing. Students were then invited to decide how its growth could be measured and then make predictions and generalizations. When this was done with one class of 12-year-old students, the results were surprising. Many produced designs and number patterns that were far more complex than anticipated (see Figure 6). Some led, for example, to quadratic patterns from which they were unable to generalize using algebra. Other lower-attaining students found it difficult to establish even an unambiguous rule for growth.

Whereas closed tasks place an upper (and lower) bound on the achievement that students might display, open tasks differentiate by response. Students are given an opportunity to show more of their true potential. While their choices may unintentionally prevent them from

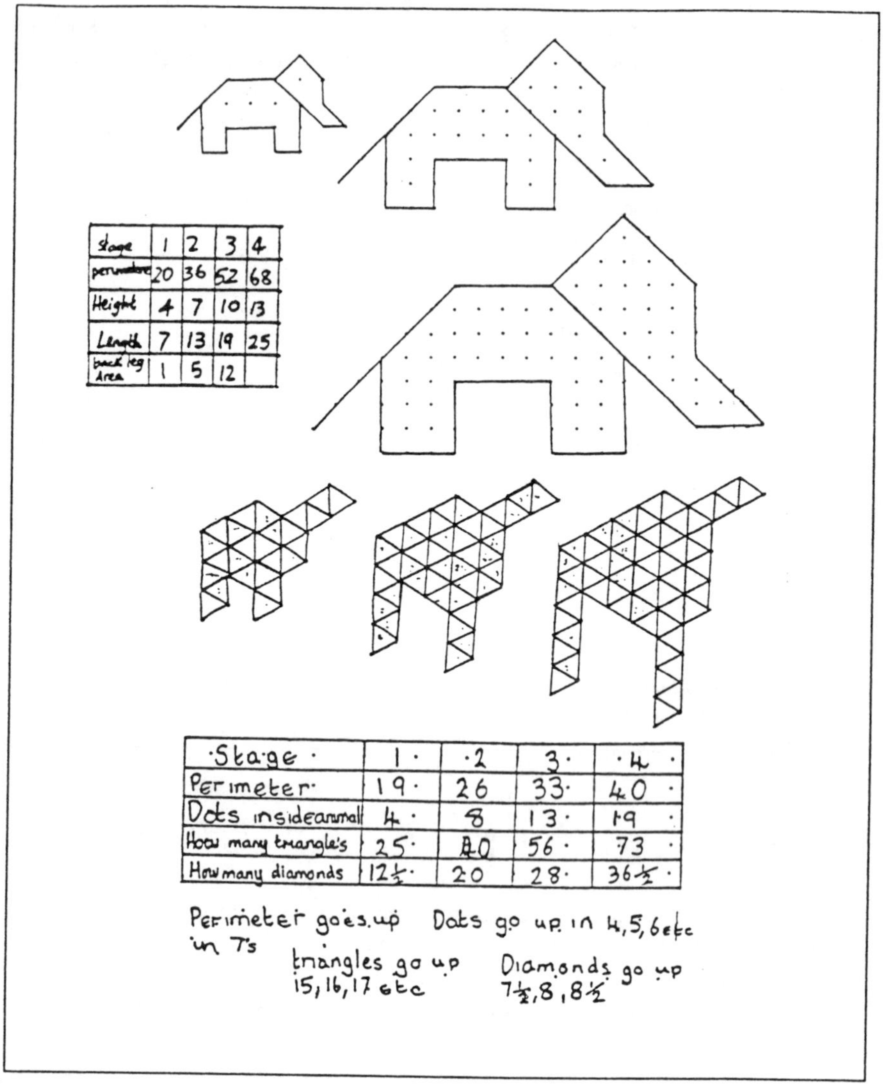

Figure 6 *Some responses to an "open" version of the number pattern task*

displaying some abilities (in the above case, algebra), they are able to display strategic skills. The marking schemes for open tasks must therefore explicitly reward such skills. Students are helped greatly if they are given the broad headings under which their work will be marked before they begin. More is said about this during the discussion of extended tasks, below.

(iv) The Degree of Autonomy and Flexibility Required

Frequently, the difficulty of a task is associated with a technical demand of the mathematics involved. It is now more widely recognized, however, that there is often a gap of several years between being able to perform a skill imitatively in a familiar context and being able to deploy that skill, perhaps in conjunction with other skills, autonomously in a non-routine situation. This has obvious assessment implications. Tasks which require a high degree of autonomy and flexibility may have to contain low technical demands if they are to be accessible to most students.

(v) The Extent and Coherence of the Task

Doing and using mathematics involve tackling a great variety of tasks, from the quick mental calculation to the extended practical problem in which different parts of mathematics may be integrated. We find it useful to distinguish tasks by length:

- *Short tasks* (which occupy anything from a few seconds to twenty minutes).
- *Long tasks* (which occupy from twenty minutes to two hours).
- *Extended tasks* (which may take many hours, and often spread over several weeks).

Traditionally, most assessment has consisted of the short task variety, but in recent years long and extended types of tasks have become a major focus of attention. This is due particularly to an increased emphasis on strategic thinking.

Whatever the overall length of the task, its *coherence* is crucial. By this we mean the extent to which a task has been broken down into smaller, shorter subtasks. It is common to find extended tasks fragmented into many small closed steps which lead the student along a predetermined path. This will obviously reduce the extent to which a student can display evidence of strategic thinking.

The success of an extended task depends largely on how much students can become self-motivated, making it "their own". This may involve students significantly in developing the task. It may need to be done in negotiation with the teacher to ensure that significant mathematical possibilities remain.

One 15-year-old girl, for example, decided to design and make a shape sorter (see Figure 7) using a "house" design. She began by sketching several designs, and evolved critiera which her design should satisfy:

Design a shape sorter

Young children are often given 'shape sorters' to help them develop eye and hand coordination.

What do you think makes a good 'Shape sorter'?
Write down your list of desirable features.

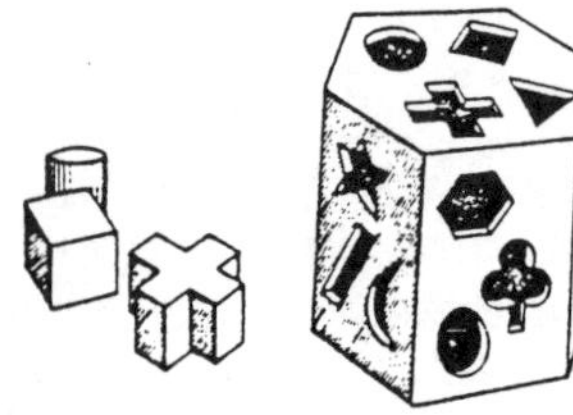

Design your own shapes and shape sorter and make them all out of cardboard.
Write about how you designed it, and why you designed it like you did.

Give your shape sorter to young children and watch them play with it.
Which shapes do they find easy to put in? Why?
Which shapes are harder? Why?

The remainder problem

When a boy counted out his sweets in fours, he had two left over;
When he counted them out in fives, he had one left over.
How many sweets did he have?
Investigate other problems like this...

Figure 7 *Two examples of extended tasks, a practical design problem, and a pure "number investigation"*

- each shape should only pass through its own slot,
- there is a simple way of retrieving shapes after they have been inserted.

After designing nets for the house and a variety of solids, including prisms based on a number of regular polygons, she made a creditable model.

Perhaps she could have gone on to consider the relative difficulty of inserting different shapes into the sorter, testing her hypotheses experimentally on a few young children.

A second student, a 13-year-old girl, decided to explore the "Remainder Problem" (Figure 7). Her work is shown in Figure 8. The particular problem under investigation may seem rather mundane, but as she attempts to generalize it, the girl's enthusiasm is captured. The concept of least common multiple is almost, but not quite, created in the course of solution; and there is the beginning of an algorithm (for the "starting numbers") when a direct explicit procedure is not available.

The Remainder Problem

The original problem was:

"When a boy counted his sweets in fours, he had two left over; when he counted them in fives, he had one left over. How many sweets did he have?"

I thought about this problem and wrote it out like this: 4r2,
6, 26, 46, 66, 86, ...
5r1

The first number of sweets he could have had was 6 but after that there were many more numbers which continued to go up in 20s. I noticed that 5×4=20 and the numbers went up in 20s.

We looked at the problems:

1. How do we find what the starting number is going to be?
2. How many do they go up in each time?
3. What happens with three or more counts?

First I tried to find what the starting number is going to be. But at first it didn't seem a very easy problem. There were exceptionally cases when 4r2 and 5r1 would add up to 6 the first number, but this rarely worked.

For a while I left this and went on to the problem of what the numbers went up in. I had found that the first numbers, when multiplied together made the amount the numbers went up in. But then I found these two cases

4r2, 4r3,
gave 22, 34, 46 gave 19, 31, 43,
6r4 6r1

This made us think that the numbers always go up either by the multiples of the two numbers or half that amount.

I thought that when all the first numbers are prime numbers the numbers go up in those numbers multiplied together, but when those numbers are not all prime numbers, the numbers go up in half this multiple. But another case cropped up and put me off the trail:

3r1,
10, 16, 22, 28, 34, 40, 46,
6r4

This time the numbers should go up in a third of 3×6

3×6 = 18 1/3 of 18 = 6

The rule seemed to be correct. If the larger number is a multiple of the smaller number, then the numbers always go up in the multiple of the two numbers divided by the smallest number.

To multiply two numbers together and divided by the smallest always leaves you with the largest number. If the numbers have no connection, they go up in the multiple of the two numbers and if one number is a multiple of the second, the numbers go up in the largest number.

Next I tried using three or four counts. Here are a few examples of them:

2r0
3r1
4r2 58, 118, 178, 238, ...
5r3

As 4 is a multiple of 2 the numbers go up in 2×3×4×5+2.

In this next example, the number have no connection so go up in 2×3×5×7 = 210

2r1
3r2
7r4 95, 305, 515, ...
5r0

I did find that there was a pattern in the tables of first numbers.

	3r0	3r1	3r2
5r0	0	10	5
5r1	6	1	11
5r2	12	7	2
5r3	3	13	8
5r4	9	4	14

	4r0	4r1	4r2	4r3
5r0	0	5	10	15
5r1	16	1	6	11
5r2	12	17	2	7
5r3	8	13	18	3
5r4	4	9	14	19

The numbers go from 0 to (in this case) 19 in order in a pattern. The numbers go from 0 diagonally down from the top corner.

	2r0	2r1	
3r0	0		
3r1		1	
3r2			□

Then you want the next number down at □ but as is goes off the table you have to look horizontally across for the 2.

Then the 3 goes off the table so you look vertically upwards from what the 3 should be. The 4 also goes off the edge of the table so you must look horizontally across to find it and then as usually diagonally down till you reach the bottom and the table is finished.

It always works in the same pattern.

numbers start here

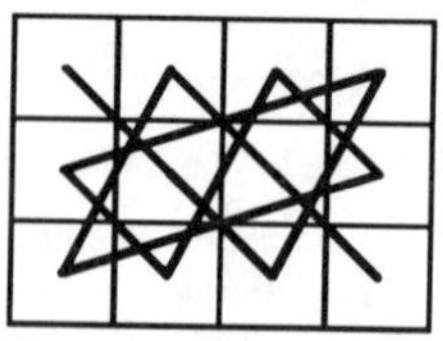

numbers finish here

This kind of table works for any numbers:

	11r0	11r1	11r2	11r3	11r4	11r5	11r6	11r7	11r8	11r9	11r10
4r0	0	12	34	36	4	16	28	40	8	20	32
4r1	33	1	13	25	37	5	17	29	41	9	21
4r2	22	34	2	14	26	38	6	18	30	42	10
4r3	11	23	35	3	15	27	39	7	19	31	43

As we know how the table works we can predict what the first number is going to be by working out from the table but not actually writing all the numbers in. If the numbers are large it would take a long time.

Conclusion
I couldn't find a really efficient way of predicting what the first number is going to be, only by the method of the tables. What I found about the way the numbers go up has been mentioned earlier. What I tried using more than 2 counts the results seemed to be the same and fitted in with any patterns or rules about two counts.

Figure 8 *Extracts from an investigation into the "Remainder Problem"*

Grading schemes for such work are necessarily of the general profile type. A number of such schemes have been developed in recent years for the assessment of coursework for the *GCSE* (*General Certificate of Secondary Education*) examination for 16-year-old students. One scheme, for example, awards marks under the broad headings: "overall design and strategy, mathematical content, accuracy, clarity of argument and presentation". Another assesses how well a student can: "understand and respond to the task, reason and make deductions, work on the task, use equipment and communicate". Under each general heading, a number of fairly crude criteria are offered, but the eventual reliability of such marking depends on the teacher's professional judgement with external moderation by the examination board to ensure comparability between schools.

Many schools have found it helpful to make students aware explicitly of such grading schemes. Indeed, they have found it helpful to encourage students to evaluate samples of extended pieces of work themselves. This kind of reflective activity can develop a student's critical faculties — an invaluable educational spin-off.

(vi) The Context in Which the Task is Set

We find it helpful to consider three categories of context:

Pure mathematical tasks, where the focus of attention is on the exploration of the structure of mathematics itself.

Applications which illustrate the use of mathematical ideas. Here the focus of the task is still on a mathematical idea, but it is embodied in a realistic (or pseudo-realistic) application. The intention is primarily to assess facility and understanding of mathematics, not to develop insights into the real world.

Real life tasks. This category is different in that the focus of attention is clearly on gaining new insights into the real world. Such tasks require the integration and deployment of mathematical and non-mathematical skills; they may contain superfluous or insufficient information and their solutions are often of real practical value.

In recent years there has been a welcome desire to introduce more applications into school examinations. The reality has been largely cosmetic. The question illustrated in Figure 9, for example, "dresses up" trigonometry in a realistic context, but it is difficult to see why anyone would want to know the angles between the legs of an ironing board. There is, however, a "real" question lurking here. Where should one place the "stops" under the top surface so that the board can be used conveniently by people of different heights?

The ironing board

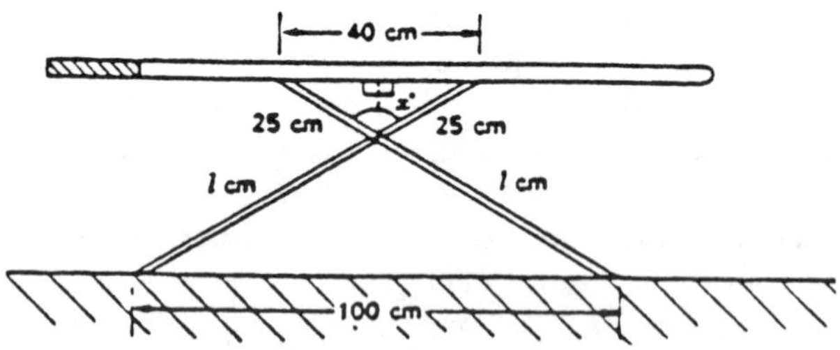

The diagram shows the side view of an ironing board.
The two legs cross at $x°$ and are equal in length.
(a) Use the information in the diagram to calculate the angle $x°$.
Give your answer to the nearest degree.
(b) Calculate the value of l.

Figure 9 *A "real" problem?*

In another long and more open question, candidates were invited to design a dog kennel from a piece of plywood of a given size. This appeared to be a "real-life" problem until one looked at the mark scheme. No credit was awarded for the practicality of the design, i.e., whether or not the design was suitable for its intended purpose. Indeed, it was quite possible to achieve full marks by designing a sealed wooden box that no dog would ever wish to inhabit! Again it becomes evident that this is a "pseudo-real" problem.

In contrast, consider the "real life" problem shown in Figure 10.

Making a wise purchase

Think of a cheap item that you often buy in the shops. (For example, breakfast cereal or coffee).
How do you go about choosing the "best" brand, quantity, etc?
Find out how other people go about choosing this item. Find out what varieties of this item are sold in different shops.
Plan and conduct an experiment to compare the quality of these different varieties.
Write a report summarizing your findings.
Say how your research will affect your purchase in the future.

Figure 10 *A "real life" problem*

This task is typical of those developed for the *Numeracy Through Problem Solving Project* (Swan, Binns & Gillespie, 1987–9). As presented here, the task does not contain sufficient support for most students or teachers, but in its original form, a support package containing detailed guidance for approximately 15-hours work, together with advice and sample tasks to assist with assessment was supplied as well. These materials have been described more fully elsewhere, e.g., Swan, Gillespie & Binns (1988). The classroom activities are divided into four stages:

Stage 1. Learning from experience. Students listen to a tape recording of a radio show which contains a number of interviews with people who have just bought different items, and an interview with two students who have been involved in producing a consumer report on choosing orange drinks. As students reflect on and analyze the tape and the report, they begin to discuss important factors and different methods of making choices.

Stage 2. Preparing for research. Students choose an item of their own to research and decide on their aims and methods.

Stage 3. Carrying out the research. Students design and conduct surveys, both inside and outside the classroom, carry out experiments and consider how they should best present their findings. This could involve posters and oral presentations in addition to written reports.

Stage 4. Presenting and evaluating the report. Written reports are now circulated around the class and any group wishing to make an oral presentation does so. Students comment on how their findings will affect their decisions in the future. The reports are evaluated by the rest of the class and each group improves its own report in the light of these comments. The final versions may be duplicated and perhaps turned into a consumer magazine for the whole school.

Each stage is accompanied by written, practical, and oral assessment tasks of different lengths, which reflect the work that students have done in small groups. Short and long tasks are provided to assess whether or not students are able to:

- identify important factors involved in decision-making,
- obtain and interpret information from oral interviews,
- obtain and interpret information from tables and graphs,
- identify possible research aims,
- select appropriate research methods,
- devise suitable methods for the collection and organization of data,
- present a summary of research in a clear and organized way,
- draw sensible conclusions from a collection of research data,
- evaluate a report and suggest improvements to it.

In addition, the complete portfolio of work assembled by each student may be assessed. This task is thus a substantial and complex activity with various conceptual and strategic objectives, each of which may be assessed by the teacher and subsequently through externally administered written assessments.

5. CONCLUDING REMARKS

If our assessment procedures are to encourage, rather than inhibit, good classroom practice then our assessment tasks should be designed to reflect those achievements that we value, not designed solely on the grounds of whether or not they are easy to mark.

We must begin, therefore, by listing those achievements and finding a range of "beautiful" tasks which reflect the balance of desired classroom activities. These may then be piloted and revised until the pupil activity matches our intentions. In the light of sample responses, a grading system may then be devised and piloted on those who will be doing the marking — external examiners, teachers, or even the pupils themselves.

The outcome will, perhaps, be a number of assessment packages comprised of tasks in a range of styles (written, oral, and practical), lengths, contexts, degrees of openness, including some which encourage autonomy and flexibility in thinking. This is a considerable design challenge, but one which we must face if we are to improve the quality of mathematical education in the classroom.

REFERENCES

Assessment of Performance Unit: 1980, *Mathematical Development: Secondary Survey Report*, no. 1, Her Majesty's Stationery Office, London.

Bell, A., Burkhardt, H. & Swan, M.: 1991, *Balanced assessment and the Mathematics Curriculum*, Shell Centre for Mathematical Education, University of Nottingham, England.

Bell, A., Shiu, C. & Horton, B.: 1981, *Evaluating Attainment in Process Aspects of Mathematics*, Shell Centre for Mathematical Education, University of Nottingham, England.

Brown, M.: 1989, 'Graded Assessment and Learning Hierarchies in Mathematics — an alternative view', *British Educational Research Journal*, **15**(2).

California Assessment Program: 1989, *A Question of Thinking, a First Look at Students' Performance on Open-ended Questions in Mathematics*, California State Department of Education, Sacramento, CA.

Committee of Inquiry into the Teaching of Mathematics in Schools: 1982, *Mathematics Counts* (the Cockcroft Report), Her Majesty's Stationery Office, London.

Department of Education and Science: 1987, *National Curriculum, Task Group on Assessment and Testing (TGAT) — A Report*, Her Majesty's Stationery Office, Department of Education and Science and the Welsh Office, London.

Graded Assessment in Mathematics: 1989, *GAIM Development Pack, Graded Assessment in Mathematics*, Macmillan, London.

Gipps, C.: 1990, *Assessment. A Teacher's Guide to the Issues.* Hodder and Stoughton, London.

Mathematical Sciences Education Board: 1989, *Everybody counts — a report to the nation on the future of mathematics education*, (NRC Mathematical Sciences Education Board), National Academy Press, Washington, DC.

Mathematical Sciences Education Board: 1990, *Assessment, A Teacher's Guide to the Issues*, Hodder and Stoughton, London.

National Curriculum Council: 1989, *Mathematics: Non-Statutory Guidance*, York.

National Curriculum Council: 1991, *Mathematics Programmes of Study, INSET for Key Stages 3 and 4*, York.

National Council of Teachers of Mathematics: 1989, *Curriculum and Evaluation Standards for School Mathematics*, Reston, VA.

Northern Examining Association: 1990, *Mathematics Through Problem Solving. A GCSE Syllabus*, Manchester.

Noss, R., Goldstein, H. & Hoyles, C.: 1989, 'Graded Assessment and Learning Hierarchies in Mathematics', *British Educational Research Journal*, **15**(2).

Swan, M., Binns, B. & Gillespie, T.: 1987–89, *Numeracy Through Problem Solving, Five modules: Design a Board Game, Produce a Quiz Show, Plan a Trip, Be a Paper Engineer, Be a Shrewd Chooser*, Shell Centre for Mathematical Education/Joint Matriculation Board, Longman, Harlow.

Swan, M., Gillespie, T. & Binns, B.: 1988, 'The assessment of numeracy through problem solving', in Pegg, J. (Ed.), *Mathematical Interfaces — Proceedings of the Australian Association of Mathematics Teachers Conference.*

Swan, M. (Ed.): 1984, *Problems with Patterns and Numbers, an examination module for secondary schools*, Shell Centre for Mathematical Education/Joint Matriculation Board, Manchester.

Swan, M.: 1985, *The Language of Functions and Graphs, An Examination Module for Secondary Schools*, Shell Centre for Mathematical Education/Joint Matriculation Board, Manchester.

Malcolm Swan
Shell Centre for Mathematical Education
University of Nottingham,
England

DEREK FOXMAN

THE ASSESSMENT OF PERFORMANCE UNIT'S MONITORING SURVEYS 1978–1987

1. INTRODUCTION AND BACKGROUND

This paper outlines the nature of the assessments and the results of national monitoring surveys carried out by the *National Foundation for Educational Research* in England and Wales on behalf of the *Assessment of Performance Unit* (*APU*) in Britain. The project began in 1977 and was closed down at the end of 1988 after carrying out six surveys of 11- and 15-year-olds in the schools of England, Wales and Northern Ireland (Scotland has its own monitoring program), the first being in 1978 and the last in 1987.

The APU was set up in 1975 at the Department of Education and Science (DES) with terms of reference which were: "To promote the development of assessing and monitoring the achievement of children at school and to seek to identify the incidence of underachievement".

Four tasks were identified relating to this remit:

- o To identify and appraise existing instruments and methods of assessment which may be relevant for these purposes;
- o To sponsor the creation of new instruments and methods of assessment, having due regard to statistical and sampling methods;
- o To promote the conduct of assessment in cooperation with Local Education Authorities (LEAs) and teachers;
- o To identify significant differences related to the circumstances in which children learn.

The APU commissioned research teams in independent research institutes in five curriculum areas to carry out these tasks. The English language, mathematics and science projects all began in 1977, although their first surveys were phased in over three years, science beginning in 1980. Foreign languages had three surveys beginning in 1981 and design and technology had one survey of 15-year-olds in 1988. Other areas were considered for monitoring: social, physical, and aesthetic development, but in the event did not materialize.

Guided by the APU's subject, technical, and other committees at the DES, the teams and their statisticians developed a wide range of assess-

M. Niss (ed.), Investigations into Assessment in Mathematics Education, 217–228.

ments, a complex variety of procedures for employing them in large-scale surveys, and various methods for analyzing the results.

The first publications were technical reports on each individual survey, with little interpretation of the results. There was some concern that the monitoring role of the APU might be unacceptable to teachers so that, initially, it was constrained by various influences on its executive committee into taking on the role of providing information for others to interpret. As the value of the results for teachers began to emerge, the APU took on a more open dissemination role which resulted in the publication of booklets and leaflets written particularly for schools.

The ages chosen for monitoring mathematics were those at the *end of primary education* for most pupils (now known as Year 6 of the National Curriculum) and the *end of compulsory secondary schooling* (Year 11). The six surveys of each age group took place annually from 1978 to 1982 and then, after a five-year interval, in 1987. After the annual series, the stated intention was to monitor every five years, but with the introduction of the *Education Reform Act* (1988) in Britain, which included the construction of a *National Curriculum and Associated Assessment* (*NCA*), the government of the day announced that the APU monitoring programme was to be terminated. The NCA was to be involved with the testing of individual students at ages 7, 11, 14, and 16 using a combination of teacher assessments and assessments developed by external agencies. Since the stated purpose of the National Curriculum is to raise standards, it seems likely that aggregated scores in the National Curriculum Assessment are intended to monitor standards, a task which was formerly included within the terms of reference of the APU.

2. THE APU MATHEMATICS ASSESSMENT FRAMEWORK

The assessment framework on which the mathematics surveys were based had *three dimensions*. It was used largely to ensure the widest possible coverage of mathematics performance.

Content. The mathematical content was covered by five main categories: number, measures, geometry, algebra and probability and statistics. Each of these categories was further divided into over a dozen sub-categories in all for detailed monitoring purposes. This division into sub-categories differed for the two age groups; number was, for example, represented by a greater number of sub-categories at Age 11, than at Age 15 while for algebra the reverse was true.

Context. The contexts in which the mathematics was placed included "everyday life", other school subjects such as science or geography, and that of mathematics itself.

Learning outcome. Three broad forms of learning outcome were assessed: The understanding of *concepts* and performance of *routine skills*; using *problem solving* strategies; and *attitudes* toward mathematics.

3. ASSESSMENT MODES

While the assessment framework remained basically the same throughout the period of the surveys, the range of modes of assessment employed was gradually extended. There were two main categories throughout, written and practical; within each there were assessments of the three main learning outcomes: concepts and skills, problem solving strategies, and attitudes. While there was some testing of problem solving, and also calculator tests for the older pupils in 1978–1982, the main emphasis in that period was on paper-and-pencil assessments of concepts and skills, and practical mathematics. In 1987, the overall balance of the assessments shifted towards more problem solving, more mathematics in context, and a greater use of technology — hand-held calculators and microcomputers. For example, in 1987 there were some written tests where students of both ages were able to use calculators if they wished, although they were not obliged to do so. There was also an experimental survey of mathematics on microcomputers. In previous years calculator skills were tested in Age 15 practical mathematics only, and microcomputers were not used.

Written Tests

Concepts and skills. Each test comprised around 50 short-response items. Only a few of these were multiple-choice questions. Normally three subcategories of content were equally represented in each test. Up to 30 such tests were used in a survey from the beginning of the series in 1978.

Although many of the questions in these tests were put in different contexts, some tests were constructed in the 1987 surveys which were built around a contextualizing theme and called *thematic tests*, e.g., "Paris Trip", "Weather", and "Garden Plan".

Problem solving strategies (problems and patterns). Five or six problems were presented in each test. A graded set of questions was presented for each. Between eight and ten tests were used in each survey. These were first included in trial form in 1980 and fully integrated into the surveys in 1981.

Attitude questionnaires. There were sections on attitudes toward mathematics in general and others on attitudes to particular topics. The scale related to the enjoyment, usefulness, and utility of mathematics, and mathematical topics. A free response section was also included with questions about pupils' experiences in their mathematical classrooms.

Practical Tests

Whereas written tests were administered by the pupils' own teachers, a feature of the practical tests was that they were given by experienced teachers of the age group concerned who were recruited and trained by the research team especially for a survey.

One-to-one practical tests. Aspects of concepts and skills, problem solving strategies, and attitudes were also assessed in the one-to-one practical tests in an oral mode by the teacher/assessors. As compared with the written tests, the emphasis was largely but not entirely on working with materials. The most important focus was on how pupils carried out the tasks. The format enabled assessors to observe and ask questions about pupils' methods and reasoning.

The training of assessors aimed to produce a high degree of standardization of presentation, but with some flexibility allowed in a friendly, but searching atmosphere. Assessors worked from a "script" containing all the questions to be put to pupils and directions on the manner in which the materials to be used should be presented. Prompts or hints were given if prescribed by the script. Flexibility was provided by the freedom given to assessors to ask for clarification of a response in a neutral way, whether scripted or not. For example, pupils could be asked how they obtained an answer whether the response was correct or incorrect. Each pupil was given about three topics from up to 15 used in a survey. Assessors recorded in as much detail as was practicable what they and the pupils said and did.

Small-group problem solving. This situation was developed for the 1987 surveys because of the growing educational interest in cooperative learning, and to include a more open-ended format in the range of assessments. Groups of three pupils of the same sex and similar attainment level in mathematics worked together on problem solving tasks with an assessor recording the activities. The tasks were problems which could be tackled in different ways and had possibilities for extension. They also had to produce discussion between the pupils so that the processes of problem solving would be more "visible" than those of pupils working silently on their own. The problems included "everyday" tasks which required planning, and more purely mathematical ones which gave opportunities for pupils to conjecture about relationships and to test out their conjecture.

An orienting task, which was interactive, was given by the assessor at the beginning of each session, but the substantive problem had to be done entirely by the students. After the pupils had done as much as they could, a further interactive session with the assessor took place for pupils to describe their findings and their opinions about the session, and for the assessor to clarify what he or she had observed or heard during it.

Mathematics with a microcomputer. Individual pupils undertook problem solving activities developed to make use of the capabilities of the microcomputer. An assessor recorded the activity on a one-to-one basis. This was an experimental probe, using a non-random sample, to illustrate the use of a micro-computer for such assessments.

4. SURVEY DESIGN

A typical structure for the 1978–1982 surveys at each age level was:

Assessment Mode	Number of Tests	Number of Pupils in Sample	Sample
Concepts and Skills	24	13,000	Whole sample
Problems and Patterns	6	1,600	Sub-sample 1
Practical 1-to-1	12	1,200	Sub-sample 2
Attitude Questionnaire	1	1,200	Sub-sample 3

Table 1 *A typical survey structure 1978–82*

The whole sample included about 2 percent of the relevant age group in England and about 6 percent of those in Wales and Northern Ireland.

In 1987 the Problems and Patterns subsample was substantially increased and there were pupils taking theme tests and calculator skills tests also. The small-group and microsamples in 1987 were drawn separately from the main sample, but all these pupils took a specially constructed Concepts and Skills test which enabled those samples to be compared with the main sample. Stratified cluster sampling was used for all random samples. All pupils in the main sample took a Concepts and Skills test, and many, in addition a second assessment. In 1987 some took three assessments.

The selection of about 50 short-response questions in a Concepts and Skills test was from between 550 and 700 used in a survey. Depending on the numbers of pupils selected for the sample in each school, all the Concepts and Skills tests in a survey were exposed in each school in the sample so that, generally, each pupil was doing a different test. There were comparable designs for other assessments except for the Attitude Questionnaire where there was only one version. This matrix sampling of the test questions enabled the overall picture to be built up by aggregating over individual contributions.

A school questionnaire gathered information about each survey school's curriculum, including topics taught, time spent on mathematics in school and on homework, size of classes, and the schools' method of organizing the classes. Other information was obtained from a computer tape of school information held by the DES and corresponding tapes from Wales and Northern Ireland. The tapes related to school type, size of relevant age group, location and pupil/teacher ratio.

5. THE DATA AND FINDINGS

Over the 12 surveys about 150,000 students took an APU Concepts and Skills test, 14,000 a Practical One-to-One test, 12,000 a Problems and Patterns test, and 17,000 an Attitude Questionnaire. Other examples of pupils tackling APU questions exist in the many hours of videotape taken for the purpose of training assessors to administer the one-to-one practical tests, and, later, the small-group assessments.

The data relating to individual questions includes not just facility (success rate) but also proportions of pupils who made specific errors and, in the case of the practical tests, proportions choosing particular materials and methods to obtain an answer.

With a data base of this size it is only possible to give a flavor of the findings in this chapter. A first point to make is that the measurement procedures did not require the tests to be the same from year to year. A lot of the reporting of content was at the item level. Although there were common questions in the surveys, a large proportion differed in successive years. New items were written and, in particular, the opportunity was taken to construct parallel questions, e.g., items which differed from each other in details such as the numbers involved, and the contextual setting or presentational features like diagrams. The effect of these features on success and error rates could be studied and so provided an element of research in the work. Parallel questions were generally not included in the same test, but were taken by equivalent samples statistically. Other features of the research were comparisons of the responses of 11- and 15-year-olds to the same items, and, within age groups, the consideration of the results by attainment band.

Concepts and Skills

An example of how pupils' understanding of decimal numbers was tracked over several surveys using these comparisons is illustrated below by the results of five multiple-choice questions. The percentage choosing each response is given and the "popular" responses described.

Question 1 (Age 15).

Which of the numbers below has the smallest value?

A	0.625	34%	(LS)
B	0.25	3%	
C	0.375	2%	
D	0.125	37%	(Correct)
E	0.5	20%	(DPI)
Other/Omit		2%	

Question 2 (Age 15).
Which of the numbers below has the smallest value?

A	0.625	4%	
B	0.25	2%	
C	0.3753	36%	(LS)
D	0.125	43%	(Correct)
E	0.5	13%	(DPI)
Other/Omit		2%	

Question 3 (Age 15).
Which of the numbers below has the largest value?

A	0.625	60%	(Correct/DPI)
B	0.25	0%	
C	0.375	0%	
D	0.125	0%	
E	0.5	33%	
Other/Omit		7%	

Question 4 (Age 11).
Sarah measured the lengths of some sticks in metres.
Which stick is the smallest?

Stick A	0.625 m	12%	(LS)
Stick B	0.25 m	4%	
Stick C	0.375 m	2%	
Stick D	0.125 m	14%	(Correct)
Stick E	0.5 m	67%	(DPI)
Other/Omit		3%	

Question 5 (Age 11).
Which of the numbers below has the smallest value?

A	0.625	25%	(LS)
B	0.25	2%	
C	0.375	1%	
D	0.125	12%	(Correct)
E	0.5	56%	(DPI)
Other/Omit		4%	

At the beginning of the 1980's we found, as did a number of researchers about this time, that decimal comparisons produced two main types of errors: one which we called the *Decimal Point Ignored* (DPI) error, where the decimals are treated as whole numbers, i.e., as if the decimal point was not there; and the second type of error is to consider that the longest decimal, or rather the largest number to the right of the decimal point, is the smallest number. This we labelled the *Largest is Smallest* error (LS).

Question 1 separates out three groups of 15-year-old pupils: those who answer correctly, and those who give the DPI or LS error responses. The responses to Question 1 illustrate also that it is the largest, not the longest decimal which the LS error makers regard as the smallest.

Question 2 shows dramatically how adding a single digit to alternative C causes the LS error makers to defect to it from alternative A.

In Question 3 the largest number instead of the smallest is asked for, and this has the effect of confounding the correct responders and those who make the DPI error – a pitfall of many decimal comparisons, of which many teachers are unaware.

Question 4 is the same as Question 1, but the percentages this time relate to the responses of 11-year-olds. The success rate of 11-year-olds is much lower than that of the older pupils and the DPI error is more prominent at Age 11 than the LS error, the reverse of the situation at Age 15. Question 5, another given to Age 11 pupils, shows yet another influence: When decimal comparisons are placed in measurement contexts there is little effect on the success rate, but many LS responders appear to revert to DPI. Interviews with pupils suggest that such pupils treat the decimal point as a divider between two different sizes of units, i.e., 0.625 meters is treated as 0 m, 625 cm. Similarly, in another item relating to time units, some pupils said that 0.7 secs meant 0 minutes, 7 seconds. In several instances there was a clear conflict between the reality of the measurements and their representation by decimal numbers. The overall effect is for the numbers to the right of the decimal point to be treated as whole numbers relating to a particular size of unit. There is a similar effect of context at Age 15.

Attainment band analyses (success and error rates within each of five 20-percent bands of attainment formed on the basis of each sample pupil's overall test score) are particularly useful in identifying the group of pupils most affected by a change in a feature of a question. We have shown that it is the lowest 20 percent attainers at both ages, who are most helped by using explanatory phrases when technical terms such as area or perimeter were otherwise used. In another example, the change from finding the cost of 4 tins of pet food at 37p a tin, to finding the cost of 9 tins, had no effect on the success rates of the top 40 percent, but had a considerable inhibiting effect on the lowest 40 percent.

Two examples of the type of information about pupils' concepts, skills, and the methods they use which were obtained from the practical tests, related to weighing and mental calculation. Asked to use a balance to find the heaviest of five wooden blocks which were identical in appearance, 91 percent of 15-year-olds did so successfully, but only 62 percent could put all five in order of weight. Most of them did so by setting aside the block already identified as heaviest and then finding the heaviest of the remainder and so on. However, as many as 10 percent of the 15-year-olds compared the heaviest block with each of the others and based their response on how far down the balance tipped each time. Pupils of both ages were asked to calculate mentally how many 18p stamps could be bought for £ (100 p = £) and then to describe their method. Eight percent of 11-year olds did this by rounding up the 18p to 20p, while 44 percent of the older pupils did so. Other methods used included successive doubling, trial multiplication, and counting up in 18's. Generally more sophisticated mental methods which were appropriate to the particular calculation were used by the older pupils. It is doubtful whether many pupils had been taught these methods.

Results such as these focus on pupils' concepts and methods, a necessary concomitant to diagnosing and responding to their needs and difficulties. The results have also revealed that cases of pupils getting right answers to questions for wrong reasons are higher than might be thought; this situation can most easily be revealed by asking them how they obtained their answers.

Problem Solving

Aspects of problem solving were assessed in both written and practical tests. The written assessments were the Problems and Patterns tests and, in 1987, there were three Practical modes: the one-to-one, Small-Group, and Mathematics on the Microcomputer. The Problems and Patterns written tests produced some interesting results, not least because, overall, girls did slightly better than boys at both ages, and by a wider margin at Age 15, a reversal of the pattern of gender differences found for nearly all the Concepts and Skills areas.

The questions in the Problems and Patterns tests were directed towards individual processes, such as generalizing a rule or systematic working, rather than open-ended problem solving. For example, in one situation pupils were given subtractions with missing figures and asked to supply all possible sets of figures which would make the subtraction correct. The first item, 5□-3□=27 (listed vertically in the actual test), had three possible sets of figures which would correctly complete the sum, while the second item had six possible sets (5□-3□=24). In the latter case the order of listing the solutions was noted: Those pupils listing them in some patterned order (e.g., 59-35; 58-34; 57-33, etc.) were regarded as systematic workers. Thirty-one percent of 11-year-olds and 56 percent of older pupils were adjudged to be systematic. Other problems which could be solved using systematic working showed that the degree to which pupils use this strategy varies with the situation.

Performance on the same mathematical problem in different contexts was looked at in the "hens and rabbits problem" where the number of heads and legs of a mixed group of these animals was given and the number of each kind of animal was asked for. In an alternative version the context was changed to one about chewing gum packets costing 2p or 4p. The difference in context led to different ways of working on the two versions: more pupils focused initially on the number of heads than on the number of legs in the animal version, while more were drawn initially to the total cost in the chewing gum version (the equivalent of the legs in the animal version) than to the number of packets.

The results of these kinds of tasks indicated the need for pupils to be familiarized with or oriented towards problem solving situations interactively. This was highlighted by contrasting pupils' attempts on one problem presented in three different modes: in the written Problems and Patterns tests, the one-to-one Practical, and the Small-Group assessment. The problem involved a number chain of whole numbers using the rule: "If it's odd add 3; if it's even, halve it". This rule results in all chains ending in either a 4–2–1 or a 6–3 loop. The format was for pupils to derive the rule by noting what happened to numbers provided by themselves and changed by the rule, to learn how to form a number chain using the rule, and to come to realize that the chains ended in two different kinds of loop. They were then given the substantive problem which was to find out what sort of numbers produced each kind of loop. The initial learning took place interactively with the assessor in the Written test mode, with a teacher assessor in the one-to-one Practical test mode, and with both a teacher assessor and by discussion among themselves in the Small-Group mode. The results were unequivocal for both age levels: More of the Small-Groups (34 percent at Age 11, 56 percent at Age 15) reached a correct conclusion than in the one-to-one Practical test (24 percent at Age 11, 28 percent at Age 15). In the Written test, only two percent of the younger and 15 percent of the older pupils got as far as indicating that multiples of 3 end in 6-3 in the written test version.

6. MONITORING: THE INDIVIDUAL AND THE SYSTEM

The APU monitoring surveys produced snapshots of performance at particular points in time. The measurement of changes in performance over time swiftly became a controversial issue as soon as the APU programme began. The catch is that the curriculum is constantly changing: Questions can go out of date and become more difficult as a result of being taught less rather than because pupils have learned less. Existing categories of performance may retain their labels but have different content — "number" for example would now contain more contextualized questions, and more calculator questions than ten years ago; a different category in fact. Another factor is that new categories of skills (e.g., investigations) enter the curriculum while others (e.g., long division) are on their way out.

Measurement models assume that the entity being measured and the measuring rod remain constant, but the curriculum does not stand still, and assessment techniques develop. There is little alternative, however, to some constancy if comparisons are to have any validity but, at the same time, it is necessary to estimate the extent to which the measure is becoming elastic over a period of time. In the 1987 Concepts and Skills tests, for example, we included a number of common items from each of the dozen or so subcategories. Those likely to be subject to obvious change such as anything about watches, digital or analogue, were not included in the comparisons. The pattern of change was interesting in that there was a distinctive improvement in geometry, and measures, and some aspects of probability/-statistics, but decline in number and algebra. The curriculum has broadened in Britain since the publication of the Cockcroft Report (CITMS, 1982), and pressure on the timetable may have resulted in such a pattern — some improvements, some declines. There was no evidence that the decline in number was due to the much greater use of calculators in the classroom, but the APU tests on calculator skills pointed up many pupil's lack of understanding of decimals and their inability to use calculators appropriately.

The APU exercise had several not always completely compatible purposes — providing a detailed overall picture of performance, finding associations between categories of performance and background variables, and monitoring over time. However, it provided a wealth of information in all these areas, providing new findings particularly in the picture of mathematical performance in Britain. The National Curriculum Assessment, which replaces it, focuses on individuals and it will be some years before its efficacy can be evaluated.

REFERENCES[1]

Committee of Inquiry into the Teaching of Mathematics in Schools: 1982, *Mathematics counts* (the Cockcroft Report), Her Majesty's Stationery Office, London.

Foxman, D., Ruddock, G., Joffe, L., Mason, K., Mitchell, P. & Sexton, B.: 1985, *A review of monitoring in mathematics 1978 to 1982: Part 1 and Part 2*, Assessment of Performance Unit, Department of Education and Science (now located at the School Examinations and Assessment Council), London.

Foxman, D., Ruddock, G., McCallum, I. & Schagen, I.: 1991, *APU Mathematics Monitoring Phase 2: 1984–1988*. Slough, NFER. School Examinations and Assessment Council, London.

Foxman, D., Ruddock, G. & McCallum, I.: 1991, *APU Mathematics Monitoring 1984–88 (Phase 2): A summary of findings, conclusions and implications*. Assessment Matters 3, School Examinations and Assessment Council, London.

Foxman, D., Hutchison, D. & Bloomfield, B.: 1991, *The APU experience 1977–1990*, School Examinations and Assessment Council, London.

1. Note on APU Bibliography. The APU produced a range of literature: review reports, reports of individual surveys, topic booklets and leaflets for teachers, articles in journals and the educational press. One volume, *The APU Experience 1977-1990*, is an account of the expertise acquired by all the APU subject teams in constructing and organizing a succession of large-scale surveys. A complete bibliography of the work in mathematics is available from Derek Foxman at the NFER, the Mere, Upton Park, Slough, Berks, SL1 2DQ, England. The main review and summary reports are listed in the references above.

Derek Foxman
National Foundation for Educational Research
Slough,
England

DAVID F. ROBITAILLE & J. STUART DONN

TIMSS: THE THIRD INTERNATIONAL MATHEMATICS AND SCIENCE STUDY

1. INTRODUCTION

The *International Association for the Evaluation of Educational Achievement* (*IEA*) is sponsoring the *Third International Mathematics and Science Study* (*TIMSS*). The major data collection phase of this study is scheduled for the 1993–94 school year, and the first international report of the study is to be published in late 1995. Over 50 countries have announced their intention to participate, and their names are listed at the end of this paper.

The International Coordinating Centre for TIMSS is housed in the Department of Mathematics and Science Education at the University of British Columbia in Vancouver, Canada. The International Coordinator for TIMSS is David Robitaille.

2. HISTORICAL BACKGROUND

IEA is well positioned to conduct international studies because of its past involvement and continuing commitment to the field of international comparative studies in education. For over 30 years, the IEA network has been setting standards for international survey research: study design, test construction tied to international curricula, scientific sampling, data analysis, and dissemination of findings to a wide international audience of policy makers, researchers, and educators. Chartered in Belgium, and with a permanent secretariat in The Hague, this non-governmental organization boasts representation from over 40 educational systems in 35 countries around the world.

IEA conducts a cycle of studies on a regular basis approximately every decade. Each of these studies is curriculum-based, issue-centered, and interdisciplinary in scope; and each survey serves as a vehicle for assessing the outputs of national education systems. IEA studies not only generate interesting univariate comparisons across individual test items and the subtests, but also permit exploration of observed differences through multivariate analyses tied to a conceptual model.

M. Niss (ed.), Investigations into Assessment in Mathematics Education, 229–244.

IEA has conducted two previous studies of mathematics and two of science. A few of the highlights of these projects are summarized in the following paragraphs.

The First International Mathematics Study

The *First International Study of Mathematics* (*FIMS*) was initiated in 1960 as IEA's first study. FIMS was essentially a comparative investigation of the outcomes of schooling with a focus on mathematics achievement as the dependent variable. The twelve countries which participated were almost all located in Europe, and were all more or less highly industrialized. FIMS examined national probability samples from two populations: 13-year-old students (Population 1) and students in their last year of secondary school (Population 2).

Almost all of the items developed for the FIMS were multiple-choice and were constructed through a collaborative international effort. The items were designed to measure student performance on various mathematics content topics at five cognitive levels: knowledge and information, techniques and skills, translation of data into symbols or vice versa, comprehension, and inventiveness. In addition to the achievement items, five attitude scales were developed along with questionnaires for students, teachers, principals, and educational experts.

FIMS produced numerous findings of interest to mathematics educators. Among the more significant were the following:

- Males outperformed females in both populations and in all countries.
- The achievement of students in Population 1 (13-year-olds) was positively associated with the level of education of the students' parents. No such association was found in Population 2.
- Population 1 students not only viewed mathematics more positively than did Population 2 students, they also rated the importance of mathematics in society more highly.
- Despite quite large differences in the mean achievement of students across countries in some instances, the achievement of the most able students in each country was very similar.

Overall, FIMS contributed to the development of a better understanding of the variability that exists across countries with respect to a number of variables with important implications for the teaching and learning of mathematics. This contribution laid the major groundwork for what would be a more intense study of the connections between what teachers do and what students learn.

The Second International Mathematics Study

The *Second International Mathematics Study* (*SIMS*) was designed to provide an international portrait of mathematics education and allowed, at every stage, for significant input and guidance from a wide range of members of the mathematics community. The students in the study were selected from two populations: Population A (students in the grade where the majority were 13.0 to 13.11 years of age by the middle of the school year), and Population B (students in their last year of secondary school and taking mathematics as a substantial part of an academic program).

Countries participated in SIMS in one or two ways. The full study was designed to provide longitudinal data comparing pretest (beginning of school year) data with posttest (end of school year) data. Countries could, however, opt to participate in only the posttest part of the study. As in FIMS, items were developed according to a content-by-cognitive-level grid. Content was divided into five strands for Population A and nine strands for Population B.

Like its predecessor, SIMS produced a large number of findings. Its emphasis on teaching practices, coupled with a closer examination of curricula across countries, generated findings which went substantially beyond those of its earlier counterpart. Some of the more important findings were that:

- There is a wide difference across countries with respect to students' opportunity to complete secondary school. Not only is this difference apparent between developed and developing countries, but it is also apparent within the group of developed countries which participated in the study.
- Classroom instruction in mathematics is largely of the "chalk-and-talk" variety with teachers relying heavily on whole-class instruction, prescribed textbooks, and common assignments.
- North American teachers engage in large amounts of review, in contrast to teachers in France and Japan (two countries whose students performed very well on the SIMS tests).
- Streaming of students into different classes on the basis of some measure of mathematical or overall ability has little effect on student achievement.
- Despite large differences in average achievement across countries, the best students generally performed at comparable levels, except in the United States. The similarity in the achievement of better students seemed to hold regardless of the retention rates in different countries.
- Large curricular differences exist across countries with respect to the concepts and content taught in both populations.

- The majority of students in all countries indicate that they believe mathematics is important, that they want to do well, and that knowledge of mathematics will be important in their careers. They also indicate that their parents share these opinions.
- Gender differences are less clear cut than in the earlier study. Females outperformed males fairly consistently on some topics and in some countries.

Overall, SIMS provided valuable information not only on the extent of growth in students' learning, but also on a variety of relationships among teaching practices, curricula, and student growth. Such information is vital in reaching a greater understanding of how mathematics learning takes place and which factors contribute to it.

The First International Science Study

The *First International Science Study* (*FISS*) was conducted in nineteen countries in 1970 as part of IEA's Six Subject Survey. Data were collected from four populations. In Population I, the modal age was 10 years; in Population II, the modal age was 14 years. Population IV was the terminal year of secondary education, with the modal age being approximately 18. The description of Population III placed it between Populations II and IV.

Some findings from the first science study were:

- Boys are more interested and do better in science in all populations, and in all countries. The differences in achievement increase with age.
- Gender is the greatest single factor affecting achievement in all countries.
- Boys are more attracted to the physical sciences, and girls are more attracted to the biological sciences. The greatest difference in achievement is in physics.
- Home background is a good predictor of achievement but there is considerable variation between countries.
- The best students do as well or nearly as well regardless of the retention rate.
- The results from the two countries which used optional tests of practical abilities, indicate that these tests measured different abilities than were measured on the standard tests.

The Second International Science Study

The *Second International Science Study* (*SISS*) was initiated in 1980, and data collection took place between 1983 and 1986. Twenty-four countries participated, including ten which had taken part in FISS. The study involved three populations of students: 10-year-olds, 14-year-olds, and students in their final year of secondary school.

Almost all of the items selected or constructed for SISS were multiple-choice. Major emphasis was placed on the sub-disciplines of biology, chemistry, and physics. Earth science topics were included, but with less emphasis. Student questionnaires were used to collect background information, data on student affective outcomes, and student perceptions of classroom practices. Teachers also completed questionnaires which sought information about background and opportunity-to-learn variables. School administrators provided data on school administrative structure and on the general context of science education.

SISS provided a great deal of information about student achievement and attitudes and about the context within which achievement and attitude development takes place. Some of the more interesting findings from SISS were:

- Levels of achievement vary widely across science disciplines within countries. For example, at the Population IV, Japan has high achievement in physics and chemistry, but relatively low achievement in biology.
- Levels of achievement vary widely between populations within countries. For example, Finnish students performed very well at Populations I and II, but very poorly at Population IV.
- Students in the highest scoring countries followed specialized programs in the sense that they were studying a very few subjects.
- Boys scored higher than girls at all levels.
- Between-school variance within countries is usually very great. There are, however, notable exceptions. Japan and Sweden, for example, show no between-school variance at the Population I level. Japan, Sweden, Norway, and Finland display low between-school variance at the Population II level.

3. VALUE OF INTERNATIONAL STUDIES

Critics of international surveys of achievement have noted that such studies require large expenditures of time, energy, and resources on the part of participants as well as investigators. In addition, questions have been raised about the significance and the impact of the findings from such studies. Insofar as the practical value to be derived from participation in international studies is concerned, Husén (1967) has described how IEA studies use the "world as an educational laboratory".

International studies provide investigators with an opportunity to assess the importance of variables which might not be available for study in their own jurisdictions: for example, class size, single-sex schools, out-of-school tutoring (Eckstein, 1982). The educational system within a given country is typically rather uniform with regard to these variables, with the result that it is not possible to manipulate those parameters to investigate the extent of their influence on variability within the system. However, by studying the influence of variables across countries, with an accompanying increase in natural variability, one can better examine the implications of alternatives. Hence, the central comparative question each country may well ask is: How do other countries educate their youth differently, and with what degree of success? Different approaches to the same goal can be compared, and their impact evaluated.

The efficacy of an educational innovation may be evaluated by examining its implementation and operation in another country. New approaches to teaching may be found by observing classroom practices of teachers from other countries. Stigler & Perry (1988) suggest that cross-cultural comparisons also lead researchers and educators to a more explicit understanding of their own implicit theories about how children learn mathematics. In all probability, this principle is generalizable to other content areas, including science. Without comparisons, we tend not to question our own traditional teaching practices and we may not even be aware of the choices we have made in constructing the educational process.

The IEA studies have reported on cross-national differences with regard to participation or retention rates, the extent of tracking or streaming, gender differences in achievement in mathematics, the content of the curriculum, and teaching practices. Many of the findings from these studies challenge the conventional wisdom which underlies educational practice, and they demonstrate that radically different approaches to the teaching and learning of mathematics and science are not only viable, but demonstrably successful. The focus for international studies should be on trying to determine the effects of particular variables on teaching and learning in the context of international comparisons. For example, it is important for North American educators to learn that 13-year-old students in France and Japan are not grouped by some measure of ability for instruction in mathematics, and yet their achievement is among the highest internationally.

A number of such important variables have been identified in international studies over the past twenty-five years. Course content, or opportunity to learn, has been shown to be an important variable in accounting for differences in achievement among students from different countries. But opportunity to learn also provides information on what is possible for students at a given age or grade level to learn.

International studies have shown that educational systems can retain very large proportions of the age cohort in the study of academic mathematics or science at the senior secondary level without penalizing the best students. Secondary analyses of the data from the first and second IEA studies in mathematics and science indicate that the highest ability students from almost all systems perform at about the same level on topics which they all have studied. The major differences among the best students appear to be mainly a function of opportunity to learn.

A multi-national, comparative study such as TIMSS differs in a fundamental way from a national study of a single system of education because it enables each country to analyze and describe its educational system in a multi-national context. The two broad questions which TIMSS seeks to address are as follows:

- How do mathematics and science educational environments differ across countries, how do student outcomes differ, and how are differences in those outcomes related to differences in mathematics and science education environments?
- Are there patterns of relationships among contexts, inputs, and outcomes internationally which can lead to improvements in the theories and practices of mathematics and science education?

Previous IEA studies in the areas of mathematics and science have produced valuable insights into a number of aspects of the teaching and learning process. In addition, they have provided a number of important lessons with respect to the design and conduct of large-scale international research projects in education. TIMSS will build on this foundation, and provide further information about a number of important matters such as:

- The need to provide current national and international *information* which countries can use to compare and contrast their curricula, teaching practices, and student outcomes with those from other countries of interest.
- An assessment of the *potential impact* that alternative curricula offerings, teaching strategies, and administrative arrangements have on learning.
- An identification of *what is possible* in the teaching of mathematics. For example, results from SIMS showed that extremely high growth rates were evident in Japan and France. That finding indicates that more significant growth rates might be possible in other countries, and that we must strive to identify the underlying causes of those high growth rates.
- A greater understanding of how and why *student attitudes* change, and what relationship the development of positive attitudes bears to classroom practices. Such an understanding is crucial to the development of a more complete picture of how mathematics learning takes place.

The earlier IEA studies have provided tantalizing glimpses of the nature of teaching and learning in mathematics and science. In addition, they have laid important foundations on which to build a deeper understanding of what takes place in the teaching and learning process.

4. ISSUES IN MATHEMATICS AND SCIENCE EDUCATION

In order to make a significant contribution to the advancement of knowledge and practice in the fields of mathematics and science education, TIMSS will address itself to issues which are of current interest and importance to a variety of stakeholders. These include teachers and other school-based personnel; supervisors at the school district, state, or national level; persons engaged in the pre-service education of teachers; educational researchers; policy makers at several levels; and parents and their elected representatives in government.

In many countries, there is considerable interest in developing workable systems of educational indicators: benchmarks against which policy makers can evaluate the current status of their school systems, establish trend lines, and report to parents and society at large about the efficacy of their schools. In some cases, such as the *OECD Indicators Project*, those efforts transcend national boundaries, and several countries cooperate to develop a common set of indicators. In the United States, the National Council of Teachers of Mathematics, which represents over 80,000 teachers and others interested in the teaching and learning of mathematics, recently published a set of *Standards* which have already begun to influence the content and the structure of the mathematics curriculum and teaching practices, at least in North America. These developments cannot help but have positive effects on students' outcomes. One of the beneficial outcomes resulting from the inclusion of science in TIMSS might well be the stimulation of similar world-wide developments in the area of science.

Similar kinds of initiatives have been or are being undertaken in other countries as well, and all of them attest to the high level of interest which exists world-wide in making available high quality and useful information on the performance of schools and students. This interest is particularly evident in science and mathematics, which are universally seen as essential components of the future success of a country's economic program. It seems likely that more countries will participate in this next round of IEA studies in science and mathematics than have ever participated in an international study before.

Specifically, TIMSS will continue the thrust of the earlier science and mathematics studies by addressing, among others, matters such as:

- international variations in the mathematics and science curricula;
- opportunity to learn;
- attitudes and opinions of students and teachers;

- students' achievement, with particular emphasis on the capability of students to apply their knowledge and skill in non-routine applications;
- the role of technology in the teaching and learning of mathematics and science;
- participation rates in college preparatory courses in mathematics and science, with particular regard to gender-based differences in rates of participation;
- practices employed by schools and school systems to direct students' course selection, including tracking and streaming;
- the nature, role, and influence of officially prescribed textbooks on the teaching of science and mathematics;
- the comparative efficacy of different approaches to the teaching of mathematics and science on student outcomes.

5. GOALS FOR TIMSS

An important objective of international studies in education is to examine the comparative influence of a variety of constituent variables within each component on the success of a given educational system as indicated by measures of students' achievement, participation, and attitudes and aspirations. The conceptual framework for TIMSS is summarized in Figure 1.

The model indicates that student outcomes in mathematics and science must be seen as embedded in a hierarchy of contexts: the local or school context, the more general contexts of mathematics and science education in a given country, and the overall societal context in which the study is conducted. Within the educational context, the *Intended Curriculum* and the *Implemented Curriculum*, which include all of the teaching and learning experiences which schools employ to foster learning and growth in their students, are crucial determinants of the quantity and quality of students' educational experiences. Within the local context, student outcomes are not limited to narrowly conceived achievement objectives which can all be measured by multiple-choice items. Instead, they include students' attitudes and opinions, higher-order thinking skills, as well as accomplishments and performance on more routine knowledge-based tasks.

The study as proposed provides participating countries with some degree of flexibility. At each age or grade level, the core study will yield comparative data for reports which will be disseminated within 14 months after the completion of data collection in 1994. In-depth analysis of findings will be reported in volumes to be published later. A variety of international options will allow subsets of countries to pursue specific topics of interest. For example, observational studies of classroom processes or of students' problem-solving strategies will allow for analyses which might help explain

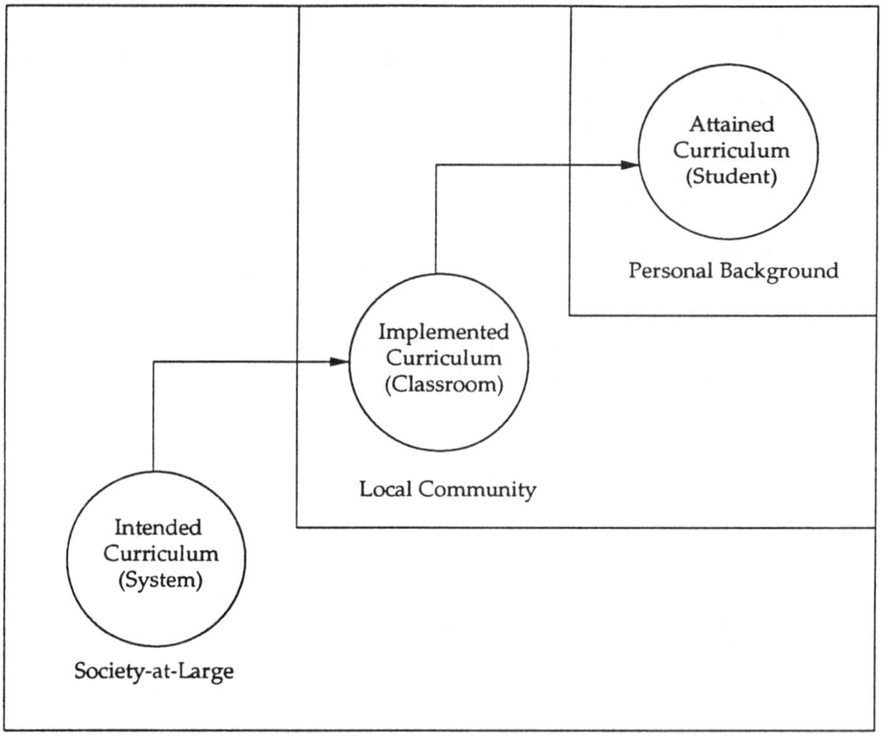

Figure 1

observed differences between classrooms and schools in different countries.

Populations

Given the primary goal of relating curriculum and teaching to achievement, the most important criterion for defining the population is when the mathematics and science being tested are taught. Population 1 will consist of students in the 2 or 3 grades that contain almost all the *9-year-olds* at the time of testing. This population was not included in SIMS, but a similar population was included in SISS and in the Reading Literacy study.

Population 2 will be drawn from students in the two or three grades which cover almost all the *students aged 13* at the time of testing. This definition is the same as for SIMS Population A; although, in SIMS Japan tested at a grade where students were somewhat younger, and some developing countries tested at a grade where many students were older. A rationale for selecting this grade-age level is that in most systems, this is about the last year when all children are still in the same type of school and are taught mathematics and science with the same curriculum. However, in many of the developing countries, a large proportion of children will no longer be in school. In other countries, the schools or the mathematics and science curricula are already strongly differentiated at this level, and that differentiation will necessitate sub-population definition and sampling.

Population 3 will consist of two sub-populations: 3a and 3b. Population 3a will include the whole cohort of students in the *last year of secondary school*, while Population 3b will consist of students who are *mathematics or science specialists*. There is a great deal of interest internationally in assessing the mathematical and scientific competency of students as they leave the secondary schools system. Every country has a vested interest in ensuring that the graduates of its school system have the mathematical and scientific literacy needed to participate fully in modern society, whether those graduates intend to study at the post-secondary level or join the work force immediately. Thus, the focus of the Population 3a study will be the mathematical and scientific competencies of the general population of students completing secondary school.

The Population 3b group would constitute a population with the same definition as was used for Population B in both SIMS and SISS. There is tremendous diversity among countries regarding the nature of the mathematics and science curricula offered to specialist students and the proportion of mathematics and science specialists in the age or grade cohort. For Population 3, testing will be cross-sectional only and take place early in the second half of the school year.

6. DESIGN OF THE STUDY

The basic study includes an in-depth curriculum analysis as well as a survey of instructional conditions, mathematical and scientific opportunities and student outcomes in Populations 1 and 2, and Populations 3a and 3b studies, focusing on school leavers. It will be designed to provide an extensive and detailed picture of the following aspects of the teaching and learning of mathematics and science internationally:

- the *intentions* for mathematics and science instruction and learning at the grade levels selected for study;
- the *organizational and social conditions* in which the instruction and learning take place;
- the *opportunities* that are presented to students for learning the content;
- the attitudes, opinions, teaching practices, and subject-matter knowledge of the *teachers*;

- the achievement, opinions, and attitudes of the *students*;
- information about *changes* in students' achievement, opinions, and attitudes during the course of a school year.

Curriculum Analysis

A major focus of TIMSS is the elaboration of the curriculum structure within each participating country. This includes both the nature of course offerings in mathematics and science, as well as the course structure in the form of programs, and a measure of opportunity to learn. The initial data collection has already occurred in some of these areas, and the methods of data collection in other aspects is at the stage of pilot trials. This concentration of effort on curricular and contextual aspects is an extension of similar foci in past IEA studies.

Two new methods have been developed to provide data for more thorough, cross-national comparison of curricula. The first, *Topic Trace Mapping* (*TTM*), is based on the curriculum frameworks developed for the mathematics and science topics of instruction. In TTM the details of the age/grade of emphasis for the areas of the curriculum, as well as the order of introduction within a topic are both reported. The encompassing nature of the curriculum frameworks, which provide the topic lists for the tracings, means that a more comprehensive analysis of curricula will be possible. In addition to determining whether or not a topic is taught in the populations of interest, the curriculum for the full grade spans of elementary and secondary school will be examined. This information will be combined with the document analysis of relevant curriculum materials to give a fuller description of the mathematics and science instruction than in past international studies.

The document analysis is a comprehensive approach to examination of the texts used for instruction. For this analysis, the text is divided into blocks, comprising units of text, and pictures or graphic material. Each block is then analyzed according to a detailed list of criteria to indicate the instructional intention of the block, including the curriculum focus. Analysis of documents in such detail will be a formidable task, but will provide a more meaningful portrait of the nature of instruction.

Data will also be collected at the system level concerning the combinations of courses that students typically take, or take to satisfy particular program requirements. This kind of information in conjunction with the detailed curriculum analysis, will provide a much more complete view of the educational systems within the countries participating in TIMSS than was possible in earlier studies.

At the level of the classroom, there are new approaches to the description of students' *Opportunity To Learn* (*OTL*). Traditionally, an OTL measure involved responses by teachers and students about particular achievement items as to whether the material required to respond to the item had been taught that year, or was assumed to have been taught in a previous year. Such measures focused on the content of the particular questions that formed the achievement test, and were thus restricted in scope. For TIMSS the nature of OTL has been expanded to include a fuller curricular range, as well as a broader view of instructional activities. Teachers will be asked about the kinds of instructional approaches they have taken to particular topics, whether they were similar to the items posed on the achievement test, or if they were of a different type. From such an extended view of the opportunities that students have had to learn specific material, a richer portrait of the instructional approaches of teachers will be gained.

Issues

In any large-scale international assessment, there are issues which need careful examination and discussion. Four aspects in particular have been the subject of review for TIMSS.

First, Wilson (1991) examined the question of validity in international studies. His premise is that the reasons for participation in such studies as TIMSS can vary widely from country to country. For some there may be interest in comparative data, while for others there may be national interests which are considered of greater importance than mere comparison. For the variety of reasons that countries participate, there is equal difficulty in predicting and producing evidence of valid instrumentation that is acceptable. As a partial response to the problem, advances in two areas are suggested. One is the trend in assessment to the incorporation of "authentic" measurement, though as Bateson, Nicol & Schroeder (1991) point out, there are conceptual aspects that require resolution rather than blind acceptance in incorporating this trend. Second, Wilson regards developments in cognition as fruitful ground both in the matter of item construction, and analysis of test performance.

A *second* critical examination of the contextual coherence of TIMSS, is discussed by Overgaard & Robitaille (1991). They elaborate the conceptual model presented in Figure 1, and discuss implications for the national, school, and teacher questionnaires, and the student attitude instruments. From an international standpoint, a more thorough understanding of the complex relationships that may be institutional or cultural in origin, and which have bearing upon interpretation of achievement results, is crucial if policy makers are to have suitable frameworks for their decisions. There is an explanatory level of comparative information beyond that of an Olympics analogy (Marklund, 1990).

The *third* issue discussed in the conceptual planning of TIMSS relates to the incorporation of alternate assessment methods. As Bateson, Nicol & Schroeder (1991) comment, the range of evaluation approaches which may be employed has implications for the construction of tables of specifications used for item development. Indeed, there is correspondence to Wilson's notion that recognition must be made of cognitive demand, as well as aspects of content and evaluation approach in constructing an appropriate specification scheme. However, as Messick (1989) and others (e.g. Resnick, 1987) note, there must be caution in the abandonment of existing measurement theory and applications, or in the uncritical acceptance of current reform suggestions. The suggestion that TIMSS "address the concerns of the authentic movement through attention to sound measurement practice, and by reference and adherence, insofar as possible, to accepted psychometric concepts, principles and language". (Bateson et al., 1991, p. 3) seems well founded.

The *fourth* area of concern focuses on participation in international studies such as TIMSS that can provide information about national educational systems which is critical for policy decisions. Such information may be of particular importance in developing countries, where determining the effectiveness of particular educational policies can be of significant benefit in planning resource allocation. It is therefore a matter of concern that the country participation list for TIMSS has significant under-representation from certain geographic areas and development levels. As Lockheed (1990) and Elfick (1990) comment, there are needs for these kinds of comparisons within and between third world and industrialized countries for appropriate educational policy review and decisions. There may be some reluctance on the part of countries to participate in cross national studies due to past experience. In many countries the cost of conducting such studies is a formidable obstacle. For some there is limited technical expertise and limited expertise for assessment, survey and policy analysis as well. For other countries there may be items on the national agenda with higher priority that participation in such international studies. Garnering the resources and support for fuller participation in international studies by developing countries is a matter the International Coordinating Centre considers important.

There are other issues which are under active consideration as planning for TIMSS proceeds. Dealing with some of these represents a continuation of the struggle with unresolved issues arising from the first two studies in mathematics and science. Dealing with others reflects a recognition of the continuing research in the areas of assessment, curriculum and survey methodology. TIMSS will undoubtedly contribute to a better understanding of some of these issues, as well as raising further research questions.

7. CONCLUSION

The activities of TIMSS represent an opportunity to provide information to education policy makers to guide education reform into the next century. Their decisions must be based on sound research, and appropriate innovations in measurement, that capture the complexity of teaching and learning in a meaningful way. The results of earlier IEA studies in mathematics and science have provided guidance for the planning of TIMSS. The continuing evolution of education throughout the world has raised particular aspects to prominence as foci of study. The potential impact of such an investment of time, effort and finances requires that all nations involved recognize the high expectations now surrounding this study, and the awesome burden it places on the researchers involved to maintain their commitment in the coming years.

REFERENCES

Bateson, D.J., Nicol, C.C. & Schroeder, T.L.: 1991, 'Alternative assessment and tables of specifications for the Third International Mathematics and Science Study', *TIMSS Working Paper*.

Eckstein, M.A.: 1982, 'Comparative school achievement', in Mitzel, H.E. (Ed.), *Encyclopedia of Educational Research*, Vol. 1, The Free Press, New York.

Elfick, J.: 1990, December, 'UNESCO remarks on TIMSS', *Proceedings of the IEA Invitational Workshop on the Third International Mathematics and Science Study*, 67–69, IEA, the Hague.

Husén, T.: 1967, *International study of achievement in mathematics*, Vols. 1 and 2, New York: Almqvist and Wiksell, Stockholm.

Lockheed, M.: 1990, December, 'Building educational research and assessment capacity', *Proceedings of the IEA Invitational Workshop on the Third International Mathematics and Science Study*, 77–91, IEA, the Hague.

Marklund, S.: 1990, December, 'IEA and educational policy', *Proceedings of the IEA Invitational workshop on the Third International Mathematics and Science Study*, 35–48, IEA, the Hague.

Messick, S.: 1989, 'Validity', in Linn, R.L. (Ed.), *Educational measurement* (3rd ed.), 13–104, Macmillan Publishing Company, New York.

Overgaard, V. & Robitaille, D.F.: 1991, 'The third international mathematics and science study: A proposed conceptual framework and research plan', *TIMSS Working Paper*.

Resnick, L.: 1987, *Education and learning to think*, National Academy Press, Washington, D.C.

Stigler, J.W. & Perry, M.: 1988, 'Cross-cultural studies of mathematics teaching and learning. Recent findings and new directions', in Grouws, D.A., Cooney, T.J. & Jones, D. (Eds.), *Effective mathematics teaching*, 194–223, National Council of Teachers of Mathematics, Reston, VA.

Wilson, R.J.: 1990, 'On the development of valid student achievement surveys for international assessments', *TIMSS Working Paper*.

Countries Participating in TIMSS

Argentina	Germany	New Zealand
Australia	Greece	Norway
Austria	Guatemala	Peru
Belgium (Flemish)	Hong Kong	Poland
Belgium (French)	Hungary	Portugal
Bulgaria	Iceland	Romania
Canada	Indonesia	Russian Federation
Chile	Iran	Singapore
China	Ireland	Slovak Republic
Colombia	Israel	Slovenia
Costa Rica	Italy	Spain
Cyprus	Japan	Sweden
Czech Republic	Korea	Switzerland
Denmark	Kuwait	Thailand
Dominican Republic	Lithuania	Tunisia
Ecuador	Mexico	USA
England/Wales	Netherlands	Zimbabwe
France		

David F. Robitaille & J. Stuart Donn
TIMSS International Coordinating Centre,
University of British Columbia,
Vancouver,
Canada

GILA HANNA

THE VALIDITY OF INTERNATIONAL PERFORMANCE COMPARISONS

1. INTRODUCTION

The *International Association for the Evaluation of Educational Assessment* (*IEA*) has conducted two comparative studies of mathematics education, first in 1967 and again in 1980. These large-scale studies, in which some 25 different countries participated, are important sources of data on student achievement, instructional practices, curricular emphases, and other indicators of the effectiveness of various educational systems. Such studies also provide information on the relative achievement of the participating countries and can serve as efficient mechanisms for monitoring international progress.

It is not easy to compare student achievement levels across countries even when the factors affecting achievement are taken into consideration. Some of these factors, such as teachers' practices, students' attitudes, societal values, and the concepts and skills acquired at various ages, are very difficult to measure. Nevertheless, because there is a high interest in international comparisons, such studies continue to be carried out despite their acknowledged limitations. In addition to the two IEA studies mentioned above, the first *International Assessment of Educational Progress* (*IAEP*) was carried out in 1988 and the Third International Mathematics and Science Study (TIMSS) has begun recently.

Differences in achievement on a test topic are normally attributed to differences in degree of knowledge and, of course, one of the purposes of such studies is precisely to obtain information about differences in knowledge among the groups tested. This is why it is so important to ensure that the tests do not contain any source of bias (any factor which would cause a consistent distortion of test results).

In comparative international studies, as in any testing situation, every effort must thus be made to ensure standardized *conditions* of test administration across classrooms and, in fact, this is relatively easy to achieve even across cultural groups. In contrast, it is very hard to ensure with different cultural groups that the *tests* themselves are equivalent. Each test must be translated into a number of different languages, and linguistic factors such as semantic variability and syntactic complexity make it difficult, if not altogether impossible, to achieve cross-cultural equivalence.

M. Niss (ed.), Investigations into Assessment in Mathematics Education, 245–252.

Even to approach such equivalence, the test must be translated very carefully, with special attention to the many peculiarities and subtleties of the various target languages. Obviously one must make sure that the meaning of a test item has not been altered in translation. But even seemingly corresponding terms in two different languages may not have the same degree of difficulty for a student. Clearly there are considerable methodological difficulties inherent in the process of translation, but if valid comparisons among countries are to be made it is crucial to ensure that test items are equivalent.

This chapter addresses *two issues* arising from the IEA *Second International Mathematics Study* (*SIMS*). The first is the cross-cultural equivalence of the original English test and its French translation as administered to the Franco-Ontarian population. The question is: *Were the students really tested with the same test*? In other words, did the translation of the test into French really preserve both the exact content of each test item and its language level? This question can be answered by having qualified people check the translation for accuracy and language level.

The second issue is that of actual bias against examinees from one of the two language groups. Even if the language in which a translated item is couched is judged to differ in its level of difficulty from that of the original, a further question remains: *Did the difference in language level affect the level of difficulty of the item itself*? This question cannot be answered definitively without empirical evidence on how different item wordings affect performance. To demonstrate that an item is biased, it must be shown (through appropriate statistical analysis) to be relatively more or less difficult for a particular sub-group than for another sub-group of the same ability.

2. DATA SOURCE

The data used in this study were drawn from the data pool of the Second International Mathematics Study (SIMS). The Population A tests (Grade 8) were administered to a random sample of 130 schools in Ontario, 115 English and 15 French. There were 180 test items divided into a core form and four rotated forms. The core form (40 items) was administered to all the students. The rotated forms (35 items each) were randomly assigned to students within a classroom, with approximately equal numbers of students responding to each form. Thus, each student answered 75 items. For technical reasons 6 of the 180 were dropped from the analysis. The remaining 174 items covered five broad topics: (1) Arithmetic (58 items), (2) Algebra (31 items), (3) Geometry (42 items), (4) Probability and Statistics (17 items), and (5) Measurement (26 items). All the items were five-alternative multiple-choice (one correct response and four distractors).

Every response to each item was coded into one of three categories: correct, wrong, or item omitted.

The test items were translated from English to French and then from the French to English. This procedure of back-translation was supposed to ensure that the translations were indeed accurate. As will be shown, however, the translations fell short of this goal.

3. METHOD AND RESULTS

Accuracy of the Translation

The 174 items were examined by six bilingual mathematics educators, each working independently. A translated item was considered not equivalent to the original English item if at least two examiners thought that it contained errors from one or more of the following six categories:

a. Differences in level of language difficulty,
b. Presence of mathematical terms that are correct translations, but nevertheless different from the terms used in the students' textbooks,
c. Differences in level of abstraction due to phrasing,
d. Differences in the intended meaning,
e. Differences in emphasis and/or clarity of notation, and
f. Minor errors in translation (e.g., typographical errors not affecting meaning).

Examples

The item reproduced below, for example, was answered correctly by 79 percent of the students responding to the English version of the test, but only by 25 percent of those responding to the French one. A look at the item reveals many differences between the two versions: In the French version there is no mention of pairs of figures; the English text makes it clear that there is one correct answer, the French does not; the English text uses both reflection and flip whereas the French text refers only to symmetry.

Item 11 in Form 3, also shown below, was answered correctly by 31 percent and 16 percent of English and French examinees respectively. In this case the meaning was changed completely in the translation. "Turn left and move one unit" is not equivalent to "On fait un pas à gauche" (which is more likely to mean simply "take a step to the left"). The correct answer to the English version is C(0,-1); the correct answer to the French one is A(1,-2).

The following are additional examples of disparities in translation:

In which diagram below is the second figure the image of the first figure under a reflection (flip) in a line?

Parmi les figures suivantes, quelles sont celles qui sont symétriques par rapport à une droite?

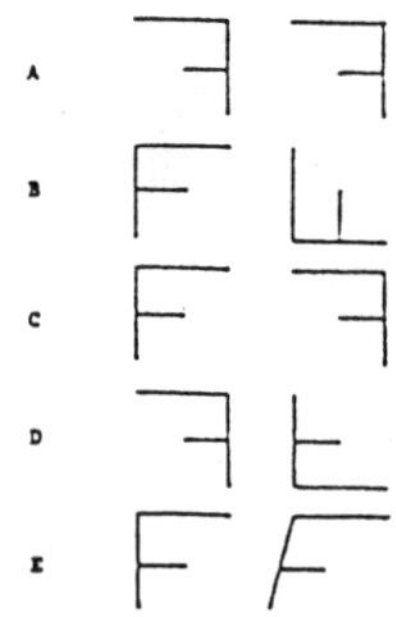

Figure 1

Suppose you start at point M(-1,-1), move a distance of one unit to N(-1,-2), then turn left and move one unit to the point P(0,-2). If you again turn left and move one unit, you will now be at the point with coordinates

Partant du point M(-1,-1), on se déplace vers le point N(-1,-2), on fait un pas à gauche et on arrive en P(0,-2). En faisant à nouveau un pas à gauche, on arrive au point de coordinnées:

A (1,-2)

B (0,-3)

C (0,-1)

D (-1,-2)

E none of these

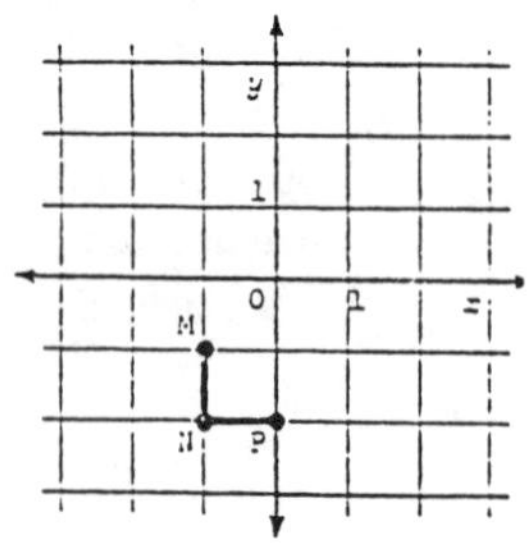

Figure 2

- Item 29 in Form 4 (error category: a, c; percent correct: English 32, French 23). The English item includes the familiar words "turn", "flip", and "slide", in addition to the technical terms "rotation", "reflection", and "translation". The French item uses technical terms only.
- Item 7 in Form 0 (error category: c; percent correct: English 70, French 42). The English item contains the explicit and detailed sentence "shows a cardboard cube which has been cut along some edges and folded out flat" versus the very formal "développement d'un cube". Also the second sentence "which two corners will touch

corner P?" is much simpler than the French "quels points seront confondus avec le point P?"

- Item 23 in Form 3 (error category: d; percent correct: English 43, French 23). Response B in French is entirely different from the English one. B ("one pair of parallel sides") is an incorrect response in the English version of the item, but a correct one in the French version ("deux côtés parallèles de longueurs égales"). Thus the item has two correct responses in French, instead of one.
- Item 27 in Form 4 (error category: e; percent correct: English 26, French 14). The variables, a, b, x, and y are in italics in English but not in French; for this reason it is difficult to understand the French item. As well, in the English version the clause "including the deposit" is underlined, whereas it is completely missing from the French version.

The above are a few striking examples of disparities. In addition, many items used relatively unfamiliar words in their French version: e.g., "consomme" versus "eaten" in Item 4 form 0, "juxtaposés" versus "put together" in Item 18 Form 2, "se transforme par rotation" versus "can be rotated (turned)" in Item 32 Form 3.

In sum, the examination by the six bilingual teachers revealed that 70 out of 174 items in the French translation (about 40 percent of the test) differed in some way from the corresponding items in English. This finding indicates that even the careful translation and back-translation of the SIMS tests had failed to achieve full equivalence of test items.

Analysis of Achievement Data

There are a number of bias-detection procedures which might be used to investigate potential bias against examinees of one language group, none without serious limitations. Among them are *Analysis of Variance* (*ANOVA*), *Transformed Item Difficulties* (*TID*), *Chi square, Item Response Theory* (*IRT*), *Mantel-Haenszel, Iterative Logit, Partial Correlation, and Distractor Response Analysis*. Each of these stems from a different conceptual perspective, purports to examine different psychometric properties of test items, and makes different assumptions about test dimensionality, item difficulty, and item discrimination. Their applicability also depends on the sample size at hand. (Unlike the other techniques mentioned above, IRT requires very large samples).

Using the core form (40 items) of the SIMS test, Simon (1987) compared the results obtained from five methods of item-bias detection (pseudo-IRT, Mantel-Haenszel, Iterative Logit, Partial Correlation, and full Chi square), and then attempted to relate these results to a subjective assessment of the linguistic equivalence of the items. Her conclusion was that although four of these statistical methods correlated highly in their identification of biased items, the biases which they identified did not seem to be related to the observed translation inaccuracies. This may indicate that there are other sources of bias in this part of the test.

In the present study, the achievement data of the two groups were examined in two different ways. The first was an investigation of the rate of omitted responses, while the second was the use of the transformed item difficulty approach to detect biased items. These two approaches were chosen because they had not been investigated by Simon (1987) and yet seemed promising.

Differences in rates of omission. A difference in rate of omitted responses may be construed as one indication that the items, originally written in English, were not adapted successfully for use with French speaking students. The mean rate of omission over the 174 items was greater for the French students than for the English students (7.1% versus 2.8%). On the average, the omission ratio of French to English was 2.5:1. The rate of omission in itself is not proof of item bias, since students omit items simply because they do not know the answer, but it is consistent with the lack of clarity observed in many of the item translations.

Table 1 shows the mean omission rates of the French and the English students by test form. The *t-tests* paired comparisons indicate that with the exception of Form 2 all the differences between the two groups are statistically significant at the .01 level.

Form	English	French	*t*-value
0	2.37	7.80	-7.18*
1	3.06	8.94	-5.89*
2	2.79	3.32	-1.13
3	2.61	6.66	-7.20*
4	3.27	8.48	-6.53*

Table 1 *Mean percent of omitted responses and t-values of differences between English and French groups*

Transformed item difficulties. Identification of biased items is problematic, since the indices available for this purpose are notoriously unreliable (Linn, Levine, Hastings & Wardrop, 1981). It has also been shown that these statistical indices of bias are essentially uncorrelated with judgement of item bias (Hoover & Kolen, 1984). An attempt was nevertheless made to explore item bias in the data through statistical means, using the technique of Transformed Item Difficulty (TID). An item is considered biased when it is comparatively more difficult to answer correctly for one language group than for the other when both language groups are of equal ability.

There was a poor correspondence between the biased items detected using the TID method and those identified as potentially biased by the judges. This may not be surprising, in view of the known limitations of the statistical method and its lack of statistical control for ability differences between the groups. Though limited in its capacity to identify specific items, the TID analysis did give an overall indication of item heterogeneity.

An examination of the TID results revealed that the items were more difficult for the French students. This in itself is not an indication of item bias, of course; it could be due to a difference in ability or curriculum between the two groups. Naturally, the statistical analysis cannot identify these external factors, they could be differences in curricular emphasis, curricular content, or translation, among others.

4. CONCLUSION

The analysis of the achievement data of English and French Ontario students showed that the inappropriate translation of certain test items may have introduced a systematic error into the measurement process, making a valid cross-language comparison of achievement very difficult if not altogether impossible.

This study has shown differences in level of difficulty between the two versions of the SIMS Grade 8 achievement tests. These differences may have been caused in part by the discrepancies between the English and French versions of the tests. Any comparison of achievement between these two Ontario language groups must take this into account.

The detection of translation biases could be assessed by direct investigation. This could be done, for example, by administering both the original and a corrected translation of the test to one group of examinees. This approach, however, would create other methodological problems, due to the repeated administration of the same test (even though the wording would be different). Another way of investigating the translation effects would be to select two groups equivalent in ability, administer to one group the original test and to the other the corrected version, and then compare the results.

It is reasonable to expect that similar differences among the other SIMS countries in the wording of their tests would also have lead to spurious differences in achievement, and thus any international comparisons of achievement should be concerned with the possibility that such disparities might be found in the other translations of this test. Clearly, bias elimination is imperative when comparing scores of students from different linguistic groups.

REFERENCES

Angoff, W.H. & Ford, S.F.: 1973, 'Item-race interaction on a test of scholastic aptitude', *Journal of Educational Measurement*, **10**, 95–105.
Hoover, H.D. & Kolen, M.J.: 1984, 'The reliability of six item bias indices', *Applied Psychological Measurement*, **8**, 173–181.
Linn, R.L., Levine, M.V., Hastings, C.N. & Wardrop, J.L.: 1981, 'Item bias in a test of reading comprehension', *Applied Psychological Measurement*, **5**, 159–173.
Simon, M.G.: 1987, *Statistical and subjective bias analyses of translated educational achievement items*, unpublished doctoral dissertation, University of Toronto.

Gila Hanna
The Ontario Institute for Studies in Education,
University of Toronto,
Canada

NORMAN L. WEBB

VISUALIZING A THEORY OF THE ASSESSMENT OF STUDENTS' KNOWLEDGE OF MATHEMATICS

1. INTRODUCTION

Assessment has begun to assume an increasingly important role in education. More than ever before national, state and district assessments are exerting pressures on teachers and students to achieve high levels of performance. National and international assessment results have drawn the attention of policy makers who use them as leverage for mandating change in education in hopes of bettering the nation's competitive edge. In the United States, the *A Nation at Risk* report (National Commission on Excellence in Education, 1983) delineated the problem by comparing American students' relatively low test scores with those of students from other industrialized nations, as well as noting the decline from 1963 to 1980 in average mathematics scores on the College Board's *Scholastic Aptitude Tests* (*SAT*).

This chapter focuses on building a theory for mathematics assessment. Several existing assumptions are challenged about assessment as it relates to mathematics education. One such assumption is that assessment is organized and delineated according to certain identifiable principles and that through understanding these principles assessment practices can be guided to specific ends.

2. THE NEEDS FOR A THEORY

Mathematical assessment is employed for a wide range of purposes — from providing information to help a teacher work with a student so that she will gain a greater understanding of number sense to plotting a national strategy that will have far reaching implications for improving mathematics education for the nation. A comprehensive theory of mathematical assessment has to be responsive to this spectrum of purposes. A theory of mathematical assessment can serve to describe, explain and predict aspects and concepts within mathematics. One direct benefit of having a theory is the clarification of terms used in association with mathematical assessment. An expanded theory would include models that give concrete structure to

M. Niss (ed.), Investigations into Assessment in Mathematics Education, 253–263.

the application of different assessment approaches and to their interaction with the outcomes.

Research in cognition and learning has tended to distinguish between domain-specific knowledge and general cognitive skills (Perkins & Salomon, 1989). There still are questions regarding the interaction between general cognitive skills and domain-specific knowledge, as well as the amount of transfer across domains. It is evident that domain-specific knowledge is important and cannot be replaced by general skills. The assessment of any domain-specific knowledge, such as the assessment of mathematics, has an important role to serve in supporting the development of this specific knowledge. However, domain-specific assessment also needs to be sensitive in detecting the relationship to general cognitive skills and the facility of individuals to use this specific knowledge in conjunction with these general cognitive skills.

The nature of mathematics itself and pedagogical approaches for teaching mathematics warrant consideration of specific assessment techniques in the area of mathematics. Deductive proof is prominent in mathematics in establishing truth, whereas the sciences depend heavily on observation and experimentation. The calculus, the algebra and the number system are axiomatic systems requiring not only a knowledge of axioms, operations and theorems, but a deep understanding of each. The power of mathematics is in its application to abstract a situation and through mathematical manipulations gain further knowledge of the phenomenon at issue. Mathematics is also dynamic and ever changing. Although calculators and computers are not unique to mathematics, their use has changed the importance of what is included in the mathematics curriculum and how it is taught. Manipulatives are assuming a greater role in mathematics classrooms to help model mathematical concepts and provide a concrete representation of students' thinking.

A theory of mathematical assessment would have to take into considerations calculators, manipulatives and the use of a variety of forms of representation. One role of such a theory would be to describe and explain differences in the assessment of mathematical knowledge with or without calculators. Such a theory would help describe why some assessment tasks are calculator sensitive while others are calculator neutral, which currently is not always evident for some tasks. Assessment techniques that have been used in other domains are being applied more and more to mathematics (Stenmark, 1989); for example, students writing responses to open-ended questions which are then scored using a holistic or primary trait method. A theory of mathematical assessment would enable educational researchers to decide whether these approaches are viable in helping to identify students' knowledge of mathematics or whether they are more an indication of writing ability. Without a theory, we must resort to lengthy efforts of exploration and trial and error in learning how to adapt assessment techniques from other domains. Even though mathematics is important in many other areas, the nature of mathematics is distinctive enough and mathematics classroom practices are different enough to suggest that mathematical assessment should be distinguishable from assessment in other content areas.

One issue that a theory of mathematical assessment should address is how the specification of the mathematical content knowledge of students to be assessed is related to the responses obtained. Another issue is the priority given in assessment to validity and reliability. In testing theory, reliability is a necessary but not sufficient condition for validity. How does this rule apply in the administration of two or three lengthy mathematical problems? A third issue is the problem of aggregation of information. Is it valid to total the number of items scored correctly on a test as an indication of a student's knowledge of mathematics? How should test scores be aggregated with teacher observations and student's projects? A fourth issue is the relationship of the assessment of mathematical knowledge to the assessment of knowledge in other content areas and to the assessment of general cognitive skills. What do they have in common and how do they differ?

3. ASSESSMENT

A number of terms refer to the collection of data and information for the purpose of describing an individual's or group's level of knowledge, performance, or achievement. The most frequently used terms are *test*, *measurement* and *assessment*. The term mathematical assessment, the focus of this chapter, is used for a particular reason. In this chapter, mathematical assessment refers to the comprehensive accounting of an individual's or group's functioning within mathematics, or in the application of mathematics. This definition is patterned after a definition of general assessment by Wood (1987), who states,

> "... assessment is regarded as providing a comprehensive account of an individual's functioning in the widest sense — drawing on a variety of evidence, qualitative as well as quantitative, and therefore going beyond the testing of cognitive skills by pencil-and-paper techniques which, for many people, is measurement ..." (p. 2).

The determination of the functioning of an individual or group within mathematics or in the application of mathematics requires considering mathematics performance in a variety of contexts, including knowledge of mathematics and disposition toward mathematics. Both qualitative and quantitative approaches are needed in order to gain a comprehensive account of this functioning, with conclusions made on the basis of information from a combination of sources.

Tests are important quantitative assessment tools, but in and of themselves do not constitute the totality of assessment. A test, defined by Cronbach (1984) as a systematic procedure for observing behavior and describing it with the aid of numerical scales of fixed categories, is a very powerful tool for mathematical assessment. The same test can be used in collecting information from one person after another. A test helps to standardize conditions so that inferences can be made about individuals and groups in a well-defined situation. This definition of a test is not restricted to paper-and-pencil techniques, but is broad enough to include questionnaires for obtaining reports on attitudes, procedures for observing behavior and observation records collected systematically such as watching students using manipulatives to demonstrate a knowledge of regrouping. However, test results are generally given as a single score or a profile of scores. It is difficult, using only numerical scores, to describe how a student draws relationships between different mathematical concepts such as a proportion and similar triangles or how a student goes about solving a problem. An expanded view of mathematical assessment embracing qualitative methods, interviews and observations, as well as teacher opinion, leads to a greater capacity to describe these important aspects of mathematical knowledge. It is this expanded view of mathematical assessment that is needed as mathematics instruction becomes directed as much toward *thinking mathematics* as *doing mathematics*.

Related to testing is *measurement*. In education, the concept of measurement comes out of the psychological testing and behaviorist tradition. "Every important outcome of education can be measured" (Ebel, 1973, p. 10). For a more precise definition, Lord & Novick (1968) call "an observable variable a measure of a theoretical construct if its expected value is presumed to increase monotonically with the construct" (p. 20). Measurement in education is restricted to a quantitative description of student behavior. As with testing, the meaning given to mathematical assessment in this chapter includes measurement as one aspect, but goes beyond the provision that the object of assessment has to be quantifiable in order to be assessed.

The concept of *assessment* must be considered from a variety of perspectives. One purpose of assessment is for it to be used as a tool by teachers to provide evidence and feedback on what students know and are able to do. At this functional level, assessment and the observations derived from its use are viewed as inherently valid and have a direct influence on teaching and learning. A second purpose of assessment is to express what is valued regarding what students are to know, do or believe. To the degree that assessment results are considered credible and important, the assessment instruments themselves may shape and influence the curriculum. In this way, assessment becomes a form of communication that sends a message from the teacher or others to the student about what it is important to know. A third purpose of assessment is to provide information to decision makers, including those within the educational system, governmental policy makers and others. At this level, assessment results are used by parents, administrators, school boards and taxpayers as the basis of judgement about the effectiveness of the educational program in general and, in some cases, about the relative skill and ability of individual teachers. A fourth purpose of assessment is to provide information on the effectiveness of the educational system as a whole. The results of certain forms of assessment — in the United States, the *SAT* (*Scholastic Aptitude Test*) and *ACT* (*American College Testing*) — have become the nation's "bottom line" indicators of the effectiveness of the educational system — regardless of the questionable validity of these conclusions. This was evidenced by the furor over the declining SAT scores from 1963 to 1980. "... the American use of tests reflects our culture's interest in quantified and 'objective' judgements, (and is a) part of the rational management ethos" (Resnick, 1981).

An analysis of the relationships between the teaching and learning of mathematics and each of the four purposes of assessment reveals the complexity of the assessment process. An evaluation of assessment must acknowledge not only its mechanics — test construction, item development and analysis — but the administrative, political and cultural influences imposed upon it as well. In practical applications, these influences can conflict and create dilemmas for teaching and learning. The pressure to produce high test scores can compete with an equal but opposing pressure to identify and address individual students' instructional needs; in some cases, for example, the evaluation of students' understanding and skills may require use of an assessment instrument on which students will perform poorly. These conflicting goals and approaches make it all the more essential that educators and policy makers have a clear understanding of the different purposes for assessment and that they evaluate the approaches to be used in light of the intended purpose.

4. SPECIFICATIONS OF CONTENT FOR MATHEMATICAL ASSESSMENT

Fundamental to mathematical assessment is an explicit definition of the content to be assessed. This definition can be derived in many ways and depends upon the purpose for the assessment, the operative conception of mathematics involved and the practical concerns of time, scoring and funding. Assessment results will vary, depending upon how the content is specified for the design of the assessment. Content specification is derived from the purpose for assessment, a conception of mathematics and a theory for the learning of mathematics. To help organize different approaches to specifying outcomes for assessment, two extreme conceptions of mathematics will be used. In reality both conceptions could be used to characterize mathematics depending on the purpose. The main point is that the assessment approach reflects an underlying conception of mathematics and will consequently influence the results from the assessment. One extreme view of mathematics is that it is a collection of facts, skills and concepts. These facts, skills and concepts can be partitioned and taught separately to students. Students who successfully master the facts, skills and concepts are considered to have a functional level of mathematics. On the other extreme mathematics is a structured body of knowledge of interdependent elements. This body of knowledge has a pattern of organization. For students to know mathematics, they need to know elements — concepts, skills, properties, principles — and the interrelationship among them. Some of the approaches used for selecting assessment tasks for each of these two conceptions of mathematics are given below.

Approaches reflecting mathematics as a collection of facts, skills and concepts

- Listing topics
- Identifying special cases
- Recording behaviors
- Monitoring performance
- Attending to processes

Approaches reflecting mathematics as a structured body of knowledge

- Choosing tasks based on statistical characteristics (e.g., factor analysis)
- Using a matrix to guide selection of tasks
- Sampling tasks from a well specified domain
- Using a conceptual rule or generating function to develop tasks
- Selecting tasks requiring an integration of knowledge

5. GENERAL COMPONENTS OF ASSESSMENT

Formulation of an assessment requires consideration of four general components of the process: the assessment *situation*, the *response* to this situation, the *analysis* of the response and the *interpretation* of the results. The interdependent nature of these components means that specification of one affects what is possible for the other three. In addition to questions of content, the manipulation of these four aspects will determine the *validity* of an assessment and the *consistency* of the results across a variety of assessment experiences. A general principle for mathematics assessment is that the situation, response, analysis and interpretation, as well as the mathematical knowledge being assessed, the characteristics of the individual or group who are to respond, and the purpose for assessment must be in alignment and agreement.

Assessment Situation

In the teaching and learning of mathematics, assessment occurs when a teacher or student gathers information about the student's knowledge, either for the purpose of determining the current state of that knowledge or for guiding further learning. An assessment situation can be defined as broadly as a test composed of a series of items or as a project extending over a period of days; as narrow as a stem to a question; — or as unobtrusively as students working on a daily assignment.

A theory for mathematics assessment needs to be useful for generating a wide range of assessment situations. Traditionally, the most common form of assessment situation has involved a test, with psychometrics as the theoretical basis (Goetz, Hall & Fetsco, in press). But, as Wainer (1986) demonstrates, the traditional "summary statistics of reliability", "item-total biserial correlation coefficients" and "mean portion correct" can be misleading. This psychometric perspective of measuring knowledge does not always coincide with assessment derived from a cognitive or information processing approach. The assessment instrument properties need to be applicable to constructing a test as to working with a student one-to-one to determine the student's thinking processes.

Response

Response in an assessment situation is what the student is expected to produce. It can vary greatly, depending on the knowledge that is required, the form of the response and whether the response is to be made by an individual or a group. The response required is related to, but not entirely dependent upon, the situation. A test item can require the student to choose the correct answer among given options, or to produce the answer as on an open-response item. The appropriateness of a particular form of response depends on the purpose of the assessment and other factors such as the time available for scoring and the number of the students being tested.

One issue involved is the degree to which the required response must be *authentic*. Asking a student to actually measure the length of an object with a ruler is more authentic than presenting the student with a drawing of a ruler and an object to be measured. Systematic observations of students doing mathematics as they work on a project and their responses to probing questions are more authentic indicators of their ability to do mathematics than a test score compiled by totaling the number of correct item responses. As noted by the *California Mathematics Council* and the *EQUALS* group (Stenmark, 1989), students' success should be evaluated while working on worthwhile investigations or tasks. One example is this performance task:

> There are 30 students in our class. The office has given us 144 pencils and 24 erasers as our supply for the year. How can we be sure we will still have pencils and erasers at the end of the school year?
>
> Are students able to make a plan?
> Can they decide when to use a calculator, and then use it effectively?
> Does everybody in the group participate?
> Do students look at all factors of the problem, or do they jump to conclusions?
> Do they recognize and use the complexities of the problem?
>
> (Stenmark, 1989, p.21)

Today, more attention is being paid to forms of assessment that deviate from the traditional forms in terms of what students are required to produce, the emphasis placed on processes and the relationship to classroom instruction. These define three major attributes of responses: the form of solution, the process as the focus, and informal classroom responses.

Analysis of the Response

A crucial step in any assessment involves the consolidation of the information collected or observed so that some meaning can be assigned in light of the identified purposes of the assessment. In the case of a test, items are traditionally scored as right or wrong, the number of items answered correctly is calculated and this total is compared to a specific standard. In classroom observations, student behaviors can be observed and recorded using a coding scheme. The marks for each behavior or action can be tabulated and the frequencies reported as they relate to other frequencies tabulated. A teacher, with some framework for depicting students' knowledge, can observe a student and then consider his or her performance in the context of this framework. This information can be used, for example, to assess whether the student is ready to learn a more complex strategy. In all of these cases, some form of analysis and abstraction of information are required in order to make judgements.

Typically, assessment results consist of a score that represents the number of correct answers. Analysis, however, involves more than simple scoring; it must consider the form of analysis, the person who does the analysis and the level of aggregation of results. Statistical analysis is not covered in this chapter because it is discussed in great detail in many sources (Lyman, 1986; Guilford, 1965; Lord & Novick, 1968). It should be noted, however, that a result can have high pedagogical validity but poor statistical qualities.

Interpretations

The final aspect of assessment involves giving meaning to the assessment outcomes and the interpretation of results. The meaning derived from assessment will depend on the generalizability of the results, the form of the reports and results, and the standards that are used for comparison of the results. Interpretation is a crucial aspect of any assessment because of the utilization of information by decision makers that goes beyond what the data or information allows. Although many forms of validity have been defined, the key question still remains: Is the assessment valid for the conclusions and interpretations that are made from the assessment?

6. TOWARD A THEORY OF MATHEMATICS ASSESSMENT

The knowledge that is currently available through the research in cognition and information processing points to the importance of knowing the structure of content in order to assess the cognitive structure of individuals related to that content. This premise, and strong arguments that educational measurement is different from psychometrics, supports the need for general theories of educational assessment that are different from classical test theory. Existing projects in mathematics education that integrate assessment with instruction testify to the need for a clear understanding of mathematical content and learning prior to assessment. The differences between the nature of mathematics and that of other subject areas lead to the conclusion that a theory of mathematics assessment should unify the principles of assessment with the unique characteristics of the structure of mathematics. Such a unification could serve as the framework for assessing mathematical knowledge and performance, both within the classroom and across large groups.

A theory of mathematics assessment must address the general issues of validity and reliability, either by accepting these concepts for the assessment of mathematics and stipulating the conditions that meet their criteria, or by arguing for their redefinition, replacement, or elimination. When assessment includes a variety of forms of information collected from different sources, for example, the concept of the convergence of information appears more applicable than the concept of reliability. However, without a theoretical basis, the true difference or similarity between convergence and reliability is difficult to decipher.

Little is known about aggregating mathematics assessment data and what form of analyses are needed to derive maximum information from one or more assessments. A theory would provide a rationale for using different aggregation procedures and a basis for determining what meaning can be assigned to the results of the various aggregated levels. The research that has been done supports the contention that valuable information can be garnered from analyzing details such as *distractor choices* on multiple-choice items, as well as from the predictive value of a total score.

A theory of mathematics assessment would have to address the ways in which mathematics assessments relate to the assessment of other forms of knowledge and educational outcomes. One relationship that would need specification is how the assessment of strategic knowledge, such as general reasoning, problem solving and planning, interacts with the assessment of mathematical knowledge. Is it possible to separate a person's knowledge of mathematics from his or her more general knowledge? Is it possible to directly assess the thinking processes a person uses in doing mathematics? Think-aloud procedures have been used to assess process, but how relevant are the results to what a student actually does? Another issue related to assessing different forms of knowledge is the question of how the assessment of mathematical *knowledge* compares with the assessment of mathematical *performance*. From a behaviorist point of view, only performance can be assessed. But cognitive theory holds that models of mental processes and structure can be constructed and used to assess the structure of knowledge. Because both performance and knowledge are relevant to the learning and teaching of mathematics, a theory must provide a basis for assessing both.

Many questions remain unanswered as we examine critically the state of mathematics assessment. Many research studies have explored different areas of assessment and thereby contributed to a better understanding of the assessment of mathematics. But there is very little mention of theory within these studies, and no mention of a theory of mathematics assessment. A theory is needed to provide a structure from which to evaluate what has been done and to direct research toward what remains to be accomplished. A theory is needed to provide a language of mathematics assessment that will further discussion and establish its relationship to teaching and learning. Time and systematic inquiry are needed to develop and test such a theory — both of which imply the need for long-range planning and directed research. The paths that culminate in a viable theory of mathematics assessment will be those grounded in a diversity of practice and those searching for unifying principles.

REFERENCES

Cronbach, L.J.: 1984, *Essentials of psychological testing*, Harper & Row, New York.

Ebel, R.L.: 1973, 'Measurement and the teacher', in Aiken, L.R., Jr. (Ed.), *Readings in psychological and educational testing*, Allyn and Bacon, Inc., Boston.

Goetz, E.T., Hall, R.J. & Fetsco, T.G.: (in press), 'Implications of cognitive psychology for assessment of academic skills', in Kamphaus, R.W. & Reynolds, C.R. (Eds.), *Handbook of psychological and educational assessment of children* (Vol. 1), Guilford Press, New York.

Guilford, J.P.: 1965, *Fundamental statistics in psychology and education*, McGraw-Hill, New York.

Lord, F.M. & Novick, M.R.: 1968, *Statistical theories of mental test scores*, Addison Wesley Publishing Company, Reading, MA.

Lyman, H.B.: 1986, *Test scores and what they mean*, Prentice-Hall, Englewood Cliffs, NJ.

National Commission on Excellence in Education: 1983, *A nation at risk: The imperative for educational reform*, U.S. Government Printing Office, Washington, DC.

Perkins, D.N. & Salomon, G.: 1989, 'Are cognitive skills context-bound?', *Educational Researcher*, **18**(1), 16–25.

Resnick, D.P.: 1981, 'Testing in America: A supportive environment', *Phi Delta Kappan*, **62**, 625–628.

Stenmark, J.K.: 1989, *Assessment alternatives in mathematics: An overview of assessment techniques that promote learning*, prepared by the EQUALS staff and the Assessment Committee of the California Mathematics Council Campaign for Mathematics, Regents, University of California, Berkeley, CA.

Wainer, H.: 1986, 'Can a test be too reliable?', *Journal of Educational Measurement*, **23**, 171–173.

Wood, R.: 1987, *Measurement and assessment in education and psychology*, The Falmer Press, London.

Norman L. Webb
Wisconsin Centre for Education Research,
University of Wisconsin-Madison,
USA

INDEX

A

B

C

D

E

F

G

H

I

P

Q

R

S